T0313849

THE
ECONOMIZATION
OF LIFE

MICHELLE MURPHY

The Economization of Life

DUKE UNIVERSITY PRESS | *Durham and London* | 2017

Library of Congress Cataloging-in-Publication Data
Names: Murphy, Michelle (Claudette Michelle), author.
Title: The economization of life / Michelle Murphy.
Description: Durham : Duke University Press, 2017. |
Includes bibliographical references and index.
Identifiers: LCCN 2016049157 (print) |
LCCN 2016051071 (ebook)
ISBN 9780822363347 (hardcover : alk. paper)
ISBN 9780822363453 (pbk. : alk. paper)
ISBN 9780822373216 (e-book)
Subjects: LCSH: Family planning — Economic aspects. |
Family policy — Economic aspects. | Fertility, Human —
Political aspects. | Fertility, Human — Social aspects. |
United States — Population policy. | Bangladesh —
Population policy.
Classification: LCC HB883.5 .M88 2017 (print) |
LCC HB883.5 (ebook) | DDC 304.6/66 — dc23
LC record available at https://lccn.loc.gov/2016049157

Cover art: Raymond Pearl's bottle of *Drosophila*
at different points in time. Pearl, *The Biology of
Population Growth*, 1930.

Contents

Acknowledgments

It is difficult to select a starting point when the work making up this book began. Perhaps in graduate school in the early 1990s when I first encountered the writing of Farida Akhter and began to think harder about what reproduction is and might be. Or in 2001 during my first year at the University of Toronto surrounded by colleagues trying to puzzle out what methods transnational feminist studies called for. Or maybe it was at the Technoscience Salon trying out ideas with Natasha Myers and so many other people. It was certainly also in the archive, following manual suction abortion studies out of feminist collectives in the United States and over to Bangladesh, an itinerary that forced me to think harder about the non-innocent transnational entanglements of reproductive technologies. This book was conceived out of all of these itineraries and more, and hence my gratitude and debts to those who have assisted this work are likewise spread in time and space beyond what I can ever do justice to acknowledge. I thank Courtney Berger at Duke University Press for her ongoing backing and intellectual companionship. This work has been funded by grants from the Social Science and Humanities Research Council, and I acknowledge not only their support but also all the labor that has gone into keeping state funding for critical humanities and social science scholarship in Canada alive through difficult times. I also thank the Jackman Humanities Institute at the University of Toronto, as well as the Clayman Institute for Gender Research at Stanford, both of which provided time and space for working on this project. I thank Farida Akhter and Unnayan Bikalper Nitinirdharoni Gobeshona (UBINIG) for their vital contributions to reproductive politics, and for making time to speak with me during my time in Bangladesh. I thank the International Centre for Diarrheal Disease Research, Bangladesh (ICDDR,B) for facilitating my research, and especially thank the library staff and the wonderful head librarian Md. Nazim Uddin for his hospitality, time, and assistance. I am grateful to the scientists and field staff at ICDDR,B as well as the Bangladesh Population Council Staff that allowed me to interview them. I thank Santi Rosario for her friendship and kindness while in Dhaka.

Many people helped with the research of this book over its long gestation, and I am indebted to the insights and responses they brought to this project: Rachel Berger, Nafisa Tanjeem, Shivrang Setler, Emily Simmonds, Sebastian Gil-Riaño, and Carla Hustak. This project is also indebted to the many opportunities to think collectively about politically charged scholarship in which I have been fortunate to take part. I thank Kavita Philip and Cori Hayden for the many conversations that molded this project in its earliest stages. I thank Adele Clarke and Vincanne Adams for our collaborations on temporality and anticipation that have continued to run through this project. I thank the members of Oxidate who have helped me to see this project through when I was stuck, and who continue to inspire me with their own work: Lochlann Jain, Jake Kosek, Jackie Orr, Elizabeth Roberts, Jonathan Metzl, Miriam Ticktin, Diane Nelson, Cori Hayden, Joseph Dumit, and Joseph Masco. I especially thank Joe Masco and Cori Hayden for generously giving their careful and brilliant responses to a full reading of the manuscript. I also thank Joseph Masco for facilitating the workshopping of the manuscript in his class at the University of Chicago, as well as the wonderful graduate students who shared their generous and incisive reactions. At the University of Toronto, I am indebted to conversations with M. Jacqui Alexander, Brian Beaton, Ritu Birla, Elspeth Brown, Dina Georgis, Deborah Cowen, Ken Kawashima, Patrick Keilty, Tong Lam, Tania Li, Sharhzad Mojab, Katharine Rankin, Sue Ruddick, Shiho Satsuka, Jon Soske, Aswhini Tambe, Judith Taylor, Malcolm Thomas, and Alissa Trotz. Thank you is also due to everyone who participates in the Technoscience Research Unit writing group. There are a multitude of conversations and scholars within the wide-ranging field of science and technology studies (STS) that have been interlocutors for this project over its long gestation: Samer Alatout, Warwick Anderson, Laura Briggs, Lisa Cartwright, Tim Choy, Joseph Dumit, Paul Edwards, Steven Epstein, Sarah Franklin, Donna Haraway, Gabrielle Hecht, Sabine Höhler, Stefan Helmreich, Chris Kelty, Martha Lampland, Hannah Landecker, Rachel Lee, Celia Lury, Maureen McNeill, Gregg Mitman, Kris Peterson, Joanna Radin, Martina Schluender, Banu Subramaniam, Lucy Suchman, Kaushik Sunder Rajan, Jennifer Terry,

Charis Thompson, Nina Wakefield, and many others. The Technoscience Salon has been a wellspring of continuous support and creativity that I have drawn on, and I thank the many people who have come and made it a place to recompose the potentials of STS scholarship. I thank the wonderful graduate students with whom I have had the privilege of thinking: Bretton Fosbrook, Justin Douglas, Kelly Ladd, Sarah Tracy, Caleb Wellum, Peter Hobbs, Duygu Kasdogan, Kira Lussier, Nicole Charles, and Brianna Hersey. I am especially thankful to my collaborator and dear friend Natasha Myers for her generosity and brilliance, as well as for her ongoing willingness to work and play together toward critical and collaborative feminist and anticolonial STS. I also thank Natasha and Astrid Schraeder for comments on early drafts during our writing group.

Early versions of some of the ideas in this book were worked out in the *Scholar and Feminist Online*, and in chapters in the collections *Relational Architectural Ecologies: Architecture, Nature, and Subjectivity*, edited by Peg Rawes; *Corpus*, edited by Paisley Currah and Monica Casper; and *Marxims and Feminisms*, edited by Shahrzad Mojab.

I am indebted to friendships that have supported me and my family, making it possible for me to take time to concentrate on writing, as well as just keep going. Thank you to Andrea Adams, Jason Brown, Kathryn Scharf, Sean Fitzpatrick, Suping Chang, Erika Iserhoff, Francis Garett, Nathaniel Price, and Suzanne Price. I thank The Commons and the Gladstone Public Library for providing our neighborhood with generative spaces to write. The online community of Academic Muse supported my focus on writing when life was busy. Heartfelt thanks to my extended family for pitching in through thick and thin. My deep gratitude for their love and care goes to Claudette Murphy, Ted Murphy, Ellen Price, and Richard Price. I thank Maceo and Mika Mercey for putting up with me as I have struggled to write, work, and parent all at once, as well as giving me so much joy and love. Matt Price has done the most to make this book possible through his skilled editorship, co-parenting, and encouragement. He has lovingly held open precious time for me to write, go into hiding, and take care of myself through the long journey of this book and all else that has happened along the way.

INTRO.1   Raymond Pearl's bottle of *Drosophila* at three points in time.
(Pearl, *The Biology of Population Growth*, 1930)

A camera captures a bottle at three points in time. It is filled with *Drosophila*, also known as fruit flies, an organism that is born, reproduces, and dies in a flicker. In the first photo, the sparsely populated bottle, rich in food, finds generations of happy fruit flies reproducing and living long lives. In the second snapshot, the busy fruit flies multiply rapidly, sharply increasing their numbers until, in the third image of the bottle, the fruit flies are so numerous the container can no longer support them, a point in time when death rises, birth declines, and population growth stagnates. The bottle becomes a container of mass death.

Looking at images of this jar today, I want to reach back, pluck open the lid, and release the fruit flies to other fates. Or I could take responsibility for feeding the flies, bred as dependent laboratory creatures by the scientific practices I care so much about. Or better yet I could smash the bottle, breaking the illusion that it is the container that conditions how the flies live or die. I want to imagine other ways of understanding aggregate life that do not demand a contained existence that ends in extermination. What would it take to smash the container?

This book is a history of two aggregate forms of life being modeled in this bottle of fruit flies: population and economy. Together population and economy have rearranged worlds over the twentieth century. New ways of valuing of life have been tied to their fates. Population and economy have been built into the architectures of nation-states where practices of quantification have helped to install economy as our collective environment, as our bottle, as our surround. How does capitalism know and dream its own conditions through numbers and data? I hope that this book will leave readers feeling and thinking differently about population and economy as adequate analytic containers for assembling life toward other futures.

Population became a new kind of experimental concern in the work of Raymond Pearl, the prominent and prolific American biologist who claimed that his 1920s experiments with fruit flies in bottles captured a law

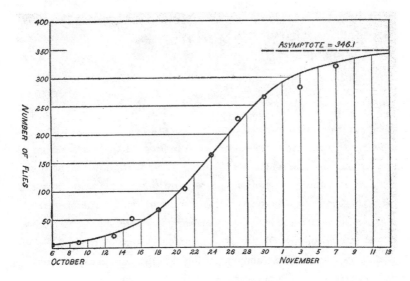

INTRO.2   Pearl's graph plotting the population growth of the fruit flies in the bottle featured in figure I.1. Pearl held that the S-curve of the line was a law of "how things grow" that reached an upper limit, or asymptote, where mass death could then outpace fertility. (Pearl, *The Biology of Population Growth*, 1930)

of "population" that governed "how things grow," and that could further be graphed as what he called "the logistic curve," today more commonly called the growth curve or the S-curve.[1] Pearl claimed this curve captured a law of life found in any aggregate of living-beings at any scale: bacteria in a petri dish, *Drosophila* in a bottle, and humans too, in a city, nation, class, or planet. The population growth curve, as a line tracing the balance of life and death in a finite container, was abstracted as a universal tendency, repeatable for all life, everywhere.[2]

Pearl promoted his work redefining "population" at the inaugural World Population Conference of 1927 held in Geneva, an event designed to propel a new international focus on problems of population that was distinct from eugenics. Organized behind the scenes by feminist birth control advocate Margaret Sanger under Pearl's supervision, the conference invited a select, mostly male, mostly American and European cohort of biologists and social scientists, along with a smattering of participants from Japan, China,

Siam, India, Argentina, Chile, Peru, and Brazil. The event promoted an expert, quantitative, and experimentalist approach to questions of "population" that critically diverged from the era's more popular eugenic orientation. Such eugenics work sought to redirect racialized heredity within evolutionary logics.[3] In the early twentieth century, eugenics had spread across the globe in projects to govern life and death toward breeding better racial futures—more fit, more pure, more evolved, more uplifted races—projects variously embraced by progressives, fascists, socialists, racists and antiracists, feminists, scientists, and political reformers.[4] Eugenics sought to manage evolutionary futures by virtue of encouraging or preventing the heredity of desirable and undesirable traits in a given population. Selective eugenic methods of directing racial futures ranged from voluntary birth control and coerced sterilization, to incarceration and segregation, to pronatalist policies and racial uplift projects, to euthanasia and mass murder. Eugenics plotted bodies, races, classes, and regions of the world on an evolutionary tree in which some bodies were more biologically progressed and forward in time (white bodies, elite bodies, male bodies, thinking bodies, able bodies), while other bodies were more primitive and pathological, and thus threatened to pull future evolution backward (colored bodies, female bodies, colonized bodies, working bodies, disabled bodies). Eugenics rested on racist claims of differential life worth based on biological difference and sought selective methods, often violent, to redirect racial futures.[5] In contrast to eugenics, at stake for Pearl in how fruit flies changed over time were not racial evolutionary futures but *economic futures*—how to balance quantitative population with national production, bringing biology and state planning together through economy.[6]

Pearl's work marks a historic shift in the status of "population" as a problematic. Pearl was trained in biometrics at the Galton Laboratory at University College London, a pivotal crossroads for both statistics and eugenics as disciplines. His work signaled a distancing from questions of racial fitness and Darwinian logics (and hence concerns with the hereditary quality of life) to an embrace of questions of *quantity* and especially the rates of birth and death within populations relative to *economic conditions*. Thus, Pearl was innovating as a biologist within a Malthusian tradition that had long tied population to political economy.[7] Importantly, his work turned "population" into an *experimental object* that could be tested and probed with the aid of fruit flies, bacteria, or chickens. Laboratory experiments could be done to populations of organisms in controlled settings. Experiments not only charted population dynamics but also sought to find ways of intervening in population's tendencies over time. Moving beyond the lab, Pearl mobilized state-produced data from censuses, as well as then emergent measures of agricultural and manufacturing production, into the project of modeling human population as yet another iteration of experiment.[8] In doing so, Pearl helped to transform "population" into a problem that needed to be both represented and intervened in at the intersection of economics and biology.

In Pearl's translation from *Drosophila* to human, the physical limits of the glass "bottle" stood in for the larger unseeable scale of "national economic production," a measure that was rapidly developing in early twentieth-century state social science. Drawing too on racialized anthropological visions of staged human progress, the purported economic container for human populations was broadly delineated as their national "stage" of economic productivity—primitive, agrarian, or mercantile, with industrial, mass-consumption capitalism as a pinnacle.[9] In contrast, the old eighteenth-century Malthusian model of population had insisted on predetermined rates of food production (the arithmetic increase of 1, 2, 3) and population growth (the geometric increase of 2, 4, 8), such that population growth would inevitably become overpopulation, unrelentingly leading to war, famine, disease, and death. Unlike Malthus, Pearl's model held that production rates were variable and *adjustable* depending on levels of civilization. Population was also adjustable as both death and birth rates could be altered with technologies and state policies. Contrary to the inevitable thrust toward crisis that concluded Malthus's law of population, Pearl's model was rife with possibilities for management.

Pearl's "proof" that the S-curve applied to humans relied on colonial data collection: the so-called natural experiment of colonized Algeria, where French colonial machinery had kept impeccable records that supposedly recorded a full growth curve.[10] According to Pearl, the "civilizing" of Algeria, and the purported improvement to agricultural productivity created by the "white man's burden" of French colonization, sparked a new "swarm" of babies, a rapidly growing aggregate of Algerians.[11] Paralleling aggregate humans with experimental insects, Pearl cited a colonial official to describe

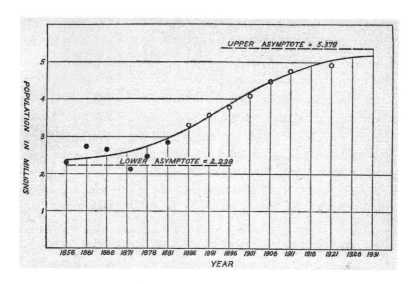

INTRO.3 Pearl's S-curve for colonized Algeria, which he claimed was a rare documented example of a human population completing a growth curve. The example was intended to mirror that of *Drosophila* in a bottle. Pearl's curve set the upper limit of the Algerian population at 5.379 million and predicted that it would stay this way unless a new set of "social, economic or other forces" came into play. (Pearl, *The Biology of Population Growth*, 1930)

the middle phase of rapid population growth, when "the natives positively pullulate under our rule" and "babies swarm among them like cockchafers under a chestnut tree in the spring."[12] Seeing Algeria as a natural petri dish, Pearl argued that as the population grew it hit a new upper limit resulting in a "process akin to natural selection [in which a] good many natives had to be eliminated before the survivors were reasonably unanimous in their belief that the old days were gone forever."[13] For Pearl, the "business of conquest" in colonized Algeria wrought the S-curve in the births and deaths of Algerians.[14] Here, the effect of the economic and colonial milieu on shaping human futures supplanted other "natural" processes.

This version of population crystallized in the period of the Cold War and decolonization into what I am calling the *economization of life*.[15] The economization of life, I argue, was and is a historically specific regime of valuation hinged to the macrological figure of national "economy." It names the practices that differentially value and govern life in terms of their ability to foster the macroeconomy of the nation-state, such as life's ability to contribute to the gross domestic product (GDP) of the nation. It is distinct from commodifying life or biocapital, or from the broader history of using quantification to monetize practices. It was not a mode that generated surplus value through labor but instead designated and managed surplus aggregate life. In this mode, value could be generated by optimizing aggregate life chances — including the reduction of future life quantity — relative to the horizon of the economy.[16] The economization of life was performed through social science practices that continued the project of racializing life — that is, dividing life into categories of more and less worthy of living, reproducing, and being human — and reinscribed race as the problem of "population" hinged to the fostering of the economy. Thus, the history of the economization of life is part of the history of racism and the technoscientific practices of demarcating human worth and exploiting life chances. Traced in this book through "population control," the economization of life was, and remains, a historically specific regime of valuation created with technoscientific practices (rather than markets) that used quantification and social science methods to calibrate and then exploit the differential worth of human life for the sake of the macrological figure of "economy."[17]

This book sketches the *epistemic infrastructures* that performed the economization of life.[18] These epistemic infrastructures were assemblages of practices of quantification and intervention conducted by multidisciplinary and multisited experts that became consolidated as extensive arrangements of research and governance within state, transnational, and nonprofit organizations. I call them *infrastructural* to underline the ways knowledge-making can install material supports into the world — such as buildings, bureaucracies, standards, forms, technologies, funding flows, affective orientations, and power relations. By attending to epistemic infrastructures, this book tracks how the experimental practices for quantifying and intervening in aggregate life consolidated into the pervasive twentieth-century infrastructures of family planning, development projects, global health, NGOs, and imperialism that were built in the name of monitoring and governing "economy" and "population." Attending to the epistemic, the book charts how "population" became a problem during a historical moment when neoliberalism was unfolding and the primary purpose of states was increasingly understood to be the fostering of "the economy," itself a historicizable twentieth-century problematic.[19] Attending to the affective, the book queries how imaginaries, feelings, futures, and phantasma are part of the work of quantification. Population and economy became massive material-semiotic-affective-infrastructural presences that can now be hard to imagine the world without.[20] They became a way for capitalism to imagine and organize its own milieu, to conjure its own conditions of possibility.

Harnessed to the enhancement of the national economy, this new era of calculative practices designated both valuable and unvaluable human lives: lives worth living, lives worth not dying, lives worthy of investment, and lives not worth being born. The history of such designations is vital for understanding how the continued racialized and sexed devaluation of life inhabits ubiquitous policies, indices, calculations, and orientations that perform new kinds of racialization even as they reject biological race as such. Moreover, this history puts questions of reproduction at the center of how capitalism summons its world.

Despite the immodesty of some of these claims, this is a short book. It is a provocation, not a proof. I have relied on the astute work of many thinkers to make these claims.[21] The book concentrates on liberal social science practices in the encounter of experts within U.S./South Asian circuits, particularly in Bangladesh. As a short book covering a century, it is dense and at the same time misses much. Compressing a century and focused on experts, calculations, and infrastructures, the book largely leaves out subtle resistances, ecologies, ethnographically rich encounters, people's variegated experiences, competing epistemologies, and local histories. Instead, I have spent years in the archive with studies, reports, and experiments produced by the transnational hegemonic project of family planning and population control. I have read these documents for the dreams and ontologies, the violence and hauntings, the counting and experiments assembled in the name of governing sex for the sake of economy. Attempting to reckon with the history of large aggregate forms (economy and population), the book ends up replicating some of the same erasures of the bird's-eye view science it studies. It tells the story of dominant structures of knowing, and risks entrenching that view. Nonetheless, I think there are virtues in telling this overarching history that come from disarticulating how phenomena that are taken for granted were installed into the world, thereby making room for other ways of thinking and being. It is a provocation and not a proof because there are many possible ways to trace the extensive history of the economization of life. Concentrating on population, liberalism, capitalism, and transnational itineraries between the United States and Bangladesh, this book traces only one of the potent routes through which the economization of life has become sedimented into the world. I invite readers to propagate the questions posed here into other itineraries, and hope that the economization of life will be further troubled.

This short book contains chapters grouped into three main parts or arcs that add up to a larger chronologically arranged account. These smaller stories are cumulative. They show the piling, propagation, and repurposing of the epistemic infrastructures that created the dense numbers and data about population for the sake of the economy. The infrastructures for experimentalizing population were built up over time, layering on top of one another, thickening the data through which differential life worth became calculable, becoming so embedded that it can it be hard to imagine the world without them.

Attending to this accumulation, this is therefore a book about reproduction in two ways. First, it thinks with and against the problem of population as a politics of reproduction commonly posed as too many births, cracking open the question of what "reproduction" is and how it might be theorized. Second, it offers a cumulative account of epistemic infrastructures reproducing themselves over time. In other words, the book asks, what is reproduced in the name of reproduction?

I came to ask questions about aggregate forms of life through my time spent in the archives of the transnational history of family planning and feminist health practices that were persistently butting against the emerging infrastructures that were governing "economy" and "population" together. While following the history of feminist practices grounded in the United States and in U.S. empire, my research was repeatedly pulled over to social science projects of the 1970s and 1980s happening in Bangladesh.[22] In that moment, Bangladesh was a crucial global node in the exuberant invention of the neoliberal practices that would make up the economization of life. Thus, my research ended up following a particular itinerary that concentrates on the circulation of practices between U.S. and Bangladeshi social science. More broadly, I argue that the economization of life was generated at this encounter between Cold War and postcolonial social science, at the crux between imperialism and decolonization, and in the tension between experiment and governance.[23]

The argument that there has been an economization of life is a grand claim, impossible to comprehensively capture, and thus this particular history serves as an opening rather than as a compendium. Arc I tells a mid-century story of U.S. Cold War quantitative practices that generated economy and population as objects of governance and intervention, tracking how figures of surplus life, and life not worth being born, became calculable. It asks how "the economy" became an affectively charged sublime, and charts how "reproduction" haunted and then later would become central to the governing of the economy. Arc II looks at the experimental exuberance of the economization of life within family planning practices of the 1960s through 1980s, with a focus on Bangladesh as a crucial site of neoliberal invention. It describes the explosion of techniques for experimental governance that sought to rearrange population through affect and counting for the sake of future economic prosperity. It shows how some (and not other) kinds of infrastructures were densely reproduced in the name of averting life and not dying, creating an era of postcolonial thick data. Arc III takes up the girling of human capital and the rise of "invest in a girl" campaigns in the 1990s that were built out of the data and numbers accumulated in the history described in arc II. It shows how the economization of life has been retwisted through financialization so that new forms of preemptive governance aspire to create real-time data about risks toward securing future supply chain logistics that stretch globally, and thus challenge the givenness of the horizon of national economy in the valuation of differential life worth. Life would explicitly become a form of capital that either increases or diminishes in value based on the riskiness of its milieu. The concluding coda reopens the question of what reproduction is, and how aggregate forms of life might be rethought through a distributed sense of reproduction. In sum, this book aspires to unsettle the world that was built, and not built, for the sake of economy and population.

Thinking with Pearl's population research is helpful for cracking open the founding relations that perpetuate within the longer history of the economization of life. There are four salient interconnected maneuvers crucial to his initial staging of the problem of population. First, his work sought to deliver a *governable* formula of population as a temporal curve plotted at the crux of aggregate life and staged economic time. As a description of change over time, and not a thing, the curve did not offer a causal explanation about how environmental and biological factors directly altered population growth (these remained open to investigation). Instead, the curve was an idealized *model* of change over time produced by experiment, which then called forth further experimental intervention: what interventions might change the curve? As an experimental instrument, the curve did not mandate any particular type of intervention but rather offered a technique that rendered legible a *target* of intervention: population growth. For example, the curve could be "smoothed" by controlling fertility or used to calculate and then encourage an "optimum population" for a given productivity.[24] Unlike Malthus's law of population, where the calamity of overpopulation in a limited world could not be avoided, with Pearl's curve mass death, famine, and overpopulation were entirely avoidable through management, as long as production "progressed" or populations were "optimized." The curve was not a mere law of nature; it was a call to action.

Second, Pearl's offering of a manipulable curve was relative to the horizon of *economic* prosperity, not improved racial kinds as in conventional eugenics. Building on a then voluminous body of eugenic statistical research on differential fertility (that is, the tendency of the poor to have more children than the rich), Pearl argued that human birth rates shifted relative to economic conditions, such that harsh, crowded, or more dangerous environments created by poverty led to higher birth rates.[25] For Pearl, shifts in birth rate relative to personal wealth "are primarily to be regarded . . . as adaptive regulatory responses—that is biological responses to evolutionary alterations in the environment in which human society lives. In this environment, the economic element is perhaps the most significant biologically."[26] Here, "economics" becomes human life's most important environmental and evolutionary correlate. The economic environment becomes the human's primary ecology. The bottle becomes the economy, rendered as the container for life, surrounding it and setting its conditions of possibility. The economic environment determined how "human units wear out faster in some occupations than others, and therefore need to be replaced faster."[27] Put more baldly, aggregate rates of fertility and death were calculable in new ways as naturalized *economic effects* in need of governing at macroscales.

Pearl's own biography tracks this shift from the qualitative preoccupations of eugenics and heredity to quantitative questions of population and production: while he began his career as a staunch eugenicist and racist, he famously made a public critique of the scientific and statistical legitimacy of much eugenic hereditary logic. Practices of eugenics, he argued, made mathematically and biologically unsubstantiated claims about the kinds of attributes that were inherited. In the early twentieth century, with over fifty years of racist evolutionary thinking to draw upon, and before DNA was agreed on as a biological mechanism of heredity, Pearl argued in the name of academic rigor (rather than antiracism) that projects to govern heredity were mathematically unscientific.[28] Pearl remained a committed racist, and continued to believe in a struggle for existence between races as they came into friction through colonialism or immigration. As an alternative to the focus on the hereditary quality of life, Pearl's work resituated the experimental study of population as a question of *economic futures* in a moment when fascists, feminists, liberals, and socialists all believed in the project of eugenics, and when the deadly force of eugenics had yet to reach its expression in European genocide. After World War II, with the retreat from eugenics, all these divergent political vectors would reattach to the problem of population in its new economic form.

Despite his critique of eugenics, Pearl's replotting of the population curve remained profoundly racist. It offered a new way to move racist accounts of differential human evolution into an economic rather than hereditary biological register. It restaged what Anne McClintock calls the asynchronic space — in which some places and bodies were plotted as more forward in time than others — that racial evolutionary logics produced into the register of economic time.[29] Now, some bodies were more forward or backward on the trajectory of economic development, or more forward or backward along the slope of the S-curve. In Pearl's model, populations with high birth rates were out of time with the forward orientation of white American economic futures.

As the economization of life continued over the twentieth century, social scientists would calculate the differential life worth of racialized bodies in terms of their contribution to future economic productivity, thereby rendering quantifiable which lives are worth being born, protected, or extended, and which lives might be abandoned or, even better, unborn. In other words, the social science practices that make up the economization of life, and that derive partially from Pearl's work, created newly legitimated quantitative ways of assigning differential life worth after explicit claims to racial biological inferiority became scientifically illegitimate. The economization of life would produce new methods of racial violence that rested on economic potential rather than bodily difference. The problem of population, as a figure of aggregate life, was replete with methods for governing brown, black, poor, and female bodies that recast racial difference in terms of economic futures. Economic futures now depended on designating overpopulation as a kind of surplus life that was better not born. Race did not have to be named in order to enact racist practices.

Third, the *Drosophila* bottles and graphical charts of Pearl's work offered a scopic regime of temporal forecasting in which individual lives are but a flicker and what comes into view are tendencies and relationships only perceivable in aggregation, at the macrodimension, across generations. More specifically, the curve abstracted out relations as *temporal rates* (rates of increase or decrease in people). In this way, the scope of the curve offered a way to *speculate* with bodies now for the sake of the future. It offered a means to make adjustments in time by acting on the future in the present. The crucial time of population was not evolutionary time but economic speculative time.

Fourth and finally, Pearl's application of the logistic curve to humans was a transnational project. It relied on data collected by cities and states, and in the case of Algeria, it relied on the census apparatus of a colonial regime and its racist commitments. Pearl's work on population was produced at the crux of race, sex, nation, colony, and metropole. Yet in the late 1920s (unlike the population control projects following decolonization), Pearl's population curve was primarily directed toward questions of governance in Europe, the United States, and Japan. His shift to an economic logic drew on recently invented measures of "national economy" concurrently forged in the United Kingdom and the United States in the birth pangs of Keynesian macroeconomics, with the corresponding invention of the measure of GDP. The elaboration of this economized rendering of "population," moreover, would continue through a transnational traffic of knowledge production and experts later moving within Cold War and postcolonial configurations. The economization of life was crafted at a threshold between a colonial and postcolonial politics of reckoning life.

With these four attributes, Pearl's curve is a harbinger of the practices that crystallized in the Cold War/postcolonial period as the "economization of life," a historically specific and polyvalent mode for knitting living-being to economy. While eugenics—oriented toward nationalist, colonial, and racial evolutionary futures—would dominate to horrific genocidal effect in the 1940s, during the decades that followed it was the epistemic practices that tied together economy and population that would flourish as both a U.S. project of foreign aid and as postcolonial projects of nation-states.[30]

This history of the economization of life complements a surge of recent scholarship, particularly in the field of science and technology studies (STS), concerned with the relations between life, reproduction, and capital. This scholarship has tracked the commodification of life in the twentieth century through patented seeds, genetic sequences, cell lines, biochemical processes, and so on, contributing to an effort to understand an emergent "politics of life."[31] In particular, feminist work has shown how central reproduction as a biological process has been to these developments, from agriculture, cloning, and clinical reproductive technologies to biotechnologies.[32] Work by Sarah Franklin, Catherine Waldby, Charis Thompson, Cori Hayden, Michael Fortun, Stephen Helmreich, Melinda Cooper, Kaushik Sunder Rajan, Kalinda Vora, and Joseph Dumit, among others, charts the rise of new speculative forms of "biowealth," "biovalue," and "biocapital," that is, the transformation of living-being (typically at micrological registers of life such as genes, molecules, viruses, algae, and cells, but also at the level of individuals, as experimental subjects in drug development, and even populations whose health is coupled to information in biobanks) into generative forms of capital through which further commodities and value are created.[33] These manifold technoscientific modes of knitting together living-being and formations of capital in the late twentieth century are accompanied, I argue, with another mode that operated through macro register or scale: the economization of life, composed of techniques for governing life for the sake of fostering "the economy," and in doing so reassembling sexed living-being at the nation-state scale of "population."

The suggestion that there is a phenomenon called the economization of life is premised on the existence of "the economy." How did "the economy" become the bottle to our fruit flies, the container for reproduction, and the surround for this late twentieth-century politics of life?

ARC I | # Phantasmagrams of Population and Economy

Worries over U.S. Economy Damped.

Mixed Feelings on Economy.

A Chilled Economy Feels a "Breath of Spring."

Why Does a "Healthy" Economy Feel So Bad?

A Jittery Economy Stirs Second Thoughts about Ostentation.

U.S. newspaper headlines have announced the collective sense of the national economy as a felt presence since the 1950s.[1] The affective force of the economy manifests as hope, optimism, jitters, confidence, panic, faith, or trepidation. It is diagnosed as healthy and ailing, sluggish or irrationally exuberant.[2] Swimming within economic tides, contemporary financiers and job seekers alike have found themselves hanging on the slightest rumor of an inside sign that might augur the economy's direction. Measures like unemployment rates, consumer confidence, and inflation all portend the economy, giving a sense of pulse and feeling for a constellation of economic activity surrounding life that one cannot hope to see in its totality. The economy becomes palpable in cryptic statements by finance ministers, a bloom of construction, or a colorful newspaper graph that can induce expectations or worries about the economy's rhythms and directions.

Given the ubiquitous force of feeling about the economy, it is easy to lose sight of the fact that "the economy," as the name for the fulsome totality of national economic activity, is an invention of the twentieth century. It was only in 1934 that the first quantitative measure of national macroeconomic output was calculated, heralding an era of econometrics that strove to explicate the dynamics and patterns that economic activities generated at the scale of the nation. Given form by quantitative models and measures, "the economy" as a noun is more than an epistemic figure. The economy became a palpable atmosphere that structured daily life, such that Americans did not need to know anything about economics, its calculations or models, in order to feel the macroeconomy as an invisible yet pressing context.[3] By 1992, "It's the economy, stupid" was a populist slogan of Bill Clinton's presidential campaign, suggesting a democratized sense of the existence of the macroeconomy. Anyone might intuitively feel the atmosphere of the economy as a determining yet diffuse presence.

John Maynard Keynes, the Cambridge economist and Bloomsbury darling, helped to bring the macroeconomy into the world in his 1936 *The General Theory of Employment, Interest and Money*, a text that catapulted him into celebrity. While European states had long tabulated the wealth of nations, the macroeconomy was not a mere counting up of wealth. Through the work of Keynes and other similarly minded macroeconomists, the national economy was explicated as a new aggregate kind, a collective blur of activity that nonetheless could be modeled as a set of predictable correlations, tendencies, forces, and rates representable in equations and graphs.[4] When interest rates go up, investment goes down, employment drops, output falls. With equations and diagrams, mathematical modeling in the 1930s performatively discerned "the economy" as a constellation of such interrelationships within a closed system whose boundary was the nation-state.[5] In turn, new projects were born to gather data about the economy, which then were congealed into measures (such as employment rates, inflation, aggregate output, consumer confidence, and so on) that would give symbolization to the well-being of the economy as a whole.[6] This constellation of relations, once legible, could then be targeted by acts of governance: decrease inflation, stimulate confidence, cut taxes. State fiscal policies would seek to alter the relationships that econometrics had brought into visibility.

Moreover, the "Keynesian revolution" in economics consisted not only in the invention of the terminology, units of analysis, and framework of macroeconomics but also in the stance that markets would not naturally correct themselves, and hence the macroeconomy not only could, but should, be governed so that capitalism might maximize employment and wealth while minimizing crashes and suffering. Within the container of "the economy," fiscal policies were aimed to control inflation, bring down unemployment, or stimulate investment, and hence were designed to trigger a cascade of interrelated effects inside the economy. Keynes's work was a direct reaction to the catastrophes of death and poverty that were World

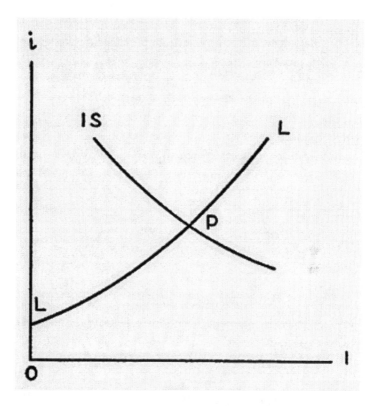

1.1   The Keynesian macroeconomic theory diagrammed as the IS/LL model, first put forward by the economist John Hicks in the journal *Econometrica* in 1937. This abstract model is based on the conversion of Keynes's theorization of the macroeconomy into an equation. It exemplifies the high levels of abstraction that went into modeling the macroeconomy as composed of interlinked and adjustable variables. The horizontal axis of this graph is national income and the vertical axis represents interest rates. The first curve, marked IS, shows how changing interest rates would affect the demand for savings. The second line, marked LL, shows how the quantity of money affects interest. This model is still taught in economics today. (Hicks, *Mr. Keynes and the 'Classics,'* 1937)

War I and the Great Depression in Britain. Macroeconomics was in this sense an attempt to develop a rationale for planning for the next war, World War II. But it was also incubated during Keynes's time as a colonial bureaucrat and applied economist in London's India Office.[7] The epistemological crafting of the macroeconomy as an atmosphere of relations that could be modeled, indexed, and governed was part of an effort to plan against poverty and prepare for war, and it was also a transference of the administrative hubris of colonial governance, which assumed a world open to interference and extraction. The macroeconomy was thus conjured, from the start, as an object to be acted on.

Through a Keynesian stance, state projects to build physical infrastructure—roads, bridges, irrigation—became a kind of fiscal policy, in which the state spent money in order to raise employment at the same time that it improved the economy's physical milieu. In so doing, Keynesian approaches argued against the automatic balancing of national budgets and instead advocated for active countercyclical fiscal policies that spent into deficit to "stimulate" the economy when it slowed. In these ways, the macroeconomy rendered capitalism as a kind of national atmosphere that needed explicating (through economics) and taking care of (through planning). Capitalism was not just made up of profit-maximizing practices within businesses and markets; through social science practices of quantification and modeling, capitalism was substantiated as a firmament of correlations, rates, and forces that surrounded life and were understood to structure conditions on the ground. Life was lived in the atmosphere of the economy.

The cloud of relationships forming the economy, moreover, was modeled as dynamic and oriented through complex looping timescapes. For Keynes, the macroeconomy was collectively shaped by the many actions that people regularly took based on their varied sense of the future, what Keynes called their "state of expectation" and "time preference."[8] From workers to financiers to state planners, Keynes posited that everyone was acting through psychological states of expectation. While most actions were decided by a sense of short-term future, Keynes held that the results of such actions nonetheless persisted into the long term, and hence governing the macroeconomy required an attention, through the aid of modeling, to such long periods. The economy was an aggregate kind that needed to be planned for the sake of the future, not just in one's own life but in generations to come.[9] Moreover, there was no natural balance of market forces that operated according to rational precepts, as neoclassical economists believed. Instead, for Keynes, the macroeconomy was probabilistic and shaped by the dizzying multitude of psychologically motivated anticipatory stances that people, businesses, and states took toward the uncertain future, stances of guessing the future that looped to then reshape that future. The temporal orientations of the macroeconomy exceeded the rhythmic ups and downs of trade cycles. The macroeconomy was an anticipatory cloud that required models to make the future palpable in the present for planners and investors alike.

For Keynes, these insights fit well with eugenics, and he considered questions of population to be within the analytic scope of macroeconomic thinking. He was a treasurer of the Cambridge University Eugenics Society, and in his 1946 Galton Lecture declared eugenics to be "the most important, significant and, I would add genuine branch of sociology that exists."[10] Keynes was against pro-natalist policies and in support of birth control, arguing that increases in prosperity were correlated to decreases in population growth, such as the decreases that he had witnessed in twentieth-century Britain. Taking the long view of macroeconomy and population, Keynes held that colonial immunization projects in India were misguided, as he believed that reductions in population caused by death from plague in Punjab were correlated with increased wages and prosperity in the generation that followed. Through aggregate and long-view scales, death from plague for Keynes became a "beneficent visitation."[11] Lives left unsaved might lead to future economic prosperity.

Importantly, the bringing into being of the macroeconomy went beyond the abstract achievement of Keynesian models. It was also the accomplishment of enormous state-sponsored efforts to gather data. In the United States and United Kingdom, respectively one rising and one soon-to-be-fading empire, the macroeconomy was brought into quantitative relief by new architectures of national accounting developed to combat the Depression.[12] Exactly how many were unemployed? How much was produced? State administrators demanded quantitative answers. Under the guidance of Nobel Prize–winning economist Simon Kuznets, the U.S. Bureau of Commerce produced its first national income account in 1934, culling data from already available tax records.[13] Soon, with war on the horizon, the demand for economic data to aid military planning intensified. What were the number of available laborers, the capacity of manufacturers to make munitions and tanks, and the minimum level of agricultural production required? It took enormous effort to produce the data behind the measures that gave shape to the macroeconomy. War machines, and not just jobs, hung in the balance.

The first calculation of annual gross national product (GNP) for the United States was produced in March 1942 in a Bureau of Foreign and Domestic Commerce report called "War Expenditures and National Production."[14] Explicitly aimed at "the problems of war production and war finance" and admittedly used as an "analytical tool, rather than as precise measurement," the GNP was an index of all goods and services produced by the government, citizens, and private enterprises, whether they were in residence or abroad, and hence was a measure amenable to the wealth of an empire.[15] In 1991 the United States was one of the last countries to switch from GNP to GDP, which measures all the goods and services produced *within* a domestic economy, irrespective of who owns the units of production, a measure amenable to an age of global capital in which the factories and companies that make up production are so often owned at a distance.[16] Of all the measures breathing form into the macroeconomy, GDP has become the most potent.

A country's GDP is revered as the best single index of "all of the economy" and celebrated as among the "great inventions of the twentieth century."[17] Economics students are taught that "GDP and related data are like beacons that help policymakers steer the economy toward the key economic objectives."[18] As an instrument, GDP takes the economy's "pulse," and thereby gauges its health and growth.[19] By 1958, the pulse of the U.S. economy was taken quarterly, giving a sense of an immediate reading of health that could then inform the "states of expectation" of bureaucrats, politicians, investors, entrepreneurs, workers, and just about anyone who follows the news. Today the pulse is taken in real time.[20]

How is GDP calculated? An economist might offer an equation and explain that GDP can be calculated in several ways.[21] One can add up all the income, or add up everything produced. These two sums are presumed to be the same, as the macroeconomy is imagined as a circulating but closed system in which all that is earned is in equilibrium with all that is spent and also all that is produced. Most textbook explanations stop there, accepting the sleight of hand that replaces "all the economy" with these particular sums. But where do these numbers come from? The source data used to calculate GDP are pulled together from other purposes: drawn from sample surveys by the Census Bureau, translated from statistics on crop yield and farming collected by the Department of Agriculture, gleaned from employment and price data surveys produced by the Bureau of Labor, or culled from income records from the Treasury.[22] The hairdresser receives a survey, fills it out, includes some payments (with receipts) and not others (paid by cash or barter), and then passes it forward; it is added with other similarly siphoned surveys that together form a sample from which to estimate the production of all hairdressers. Data are pulled from this host of sample surveys and then used to create estimates that are combined to extrapolate the fulsomeness of GDP.

Thus, the tabulation of GDP requires a large infrastructure of data collection and sampling to be in place, a state apparatus with many tentacles drawing data into its maw. The macroeconomy needs a steady stream of thick data to inform it. Each kind of data has a "vintage."[23] That is, there is a lag between the moment the data measures and its arrival into the national accounts. Moreover, the more processed the data, the more reliable and older they get. Fresher data are more uncertain. This makes GDP an estimate always under revision, with tweaked calculations regularly updating GDP in retrospect.

So while quarterly GDP is treated as a snapshot picture of the economy as a whole, it is simultaneously accepted by economists as a fuzzy number, always an estimate made from other estimates generated out of the imperfect data of a moving phenomenon. Economists know, then, that GDP can never fully count the economy. Everyone is in agreement that GDP only performatively and provisionally measures the pulse. Though GDP is dense with data, as a single number it must necessarily misrepresent. Yet all the labor and infrastructure, as well as all the guesses and gaps, behind producing GDP are rarely publicly discussed. Despite all the excitations it evokes, GDP is understood to be a single revelation of the greater sublime that it merely touches. The micromovements of GDP are watched with feeling precisely because it takes the pulse of a greater omnipresence. It is the beloved sign of the well-being of our economic habitat to which everything from state practices to the actions of hairdressers to the food we eat are attached. What won't be done, what is not open to rearrangement, for the sake of taking care of GDP?

In other words, we came to live in a world that shimmers with economic forces brought into relief through practices of quantification that do more than just aggregate, measure, and model with number.[24] The macroeconomy is miraculated for us through measures like GDP that operate as *phantasmagrams*, quantitative practices that are enriched with affect, propagate imaginaries, lure feeling, and hence have supernatural effects in surplus of their rational precepts. The term *phantasmagram* draws attention to the felt and astral consequences of social science quantitative practices, such as algorithms, equations, measures, forecasts, models, simulations, and cascading correlations.[25] I use the term to name the affective stimulus of an index like GDP beyond its calculative efficiencies. Phantasmagrams like GDP posit a fulsome beyond of immaterial forces that they explicate. At the same time, GDP is invested with aspirations and worries tied to the promises of planned intervention and the threat of future outcomes. While many economists factor in desire, pleasure, happiness, and psychological states of individuals in their explications, I am trying to point to an extra-subjective affective dimension to quantification.[26] Phantasmagrams conjure ineffable realms that can take shape as a collective phantasy in excess of the representational and logical limits of quantification practices themselves.[27]

The GDP has come to inhabit just such a collective imaginary, helping people feel the excited or sluggish condition of the economy. Beyond accounting, it is party to a mass enchantment aided by infrastructures of data-making and analysis, a jointly felt economic haze in surfeit of what any individual dreams or desires alone. As an artifact of a technical and epistemic infrastructure, GDP contributes to a nonindividual and nonconscious phantasy that substantiates the world. It points to the numinous quality of numbers. The economy, in all its enormity and detail, filled with time loops and accommodating every transaction, exceeds thought, and escapes counting (except perhaps in an alternate science fiction technocratic universe that is able to see every monetary exchange, every gesture of a labor, every item added to the forces of production), and yet it is nonetheless palpably present. The economy, uncountable and yet felt, infiltrates the sense of the world. Almost as soon as the United States calculates its GDP, so too does

India. The sublime of the economy is glimpsed while scanning help-wanted ads, in the flash of advertising that decorates urban walls, in the scavenging for a day's wage, in the posters festooning economic development initiatives, the lightness of a purse when rent is due, or when the debt collectors come for their payment. The economy pulsates in the everyday; it is worried over and cared about, beyond the number crunching and models performed by experts. The GDP no longer merely explicates the economy into measurable epistemological existence. Such measures conjure a greater dissipated canopy of forces whose conspicuous existence exceeds capacities to calculate. Economy is capitalism's secular divine and GDP its oracle. It demands faith even if on earth its worldly manifestations come in the form of blight and hype. What won't we do for GDP?

The concept of phantasmagram, as a description of GDP, draws attention to the affectively charged and extraobjective relations that are part of the force of numbers. Phantasma accompany quantification as another kind of output, as the felt, aspirational, and consequential imaginaries that structure the world in profound ways, as a potent aura in surfeit of facticity. A commodity is more than just a thing for sale made of labor; it is also an object of desire, and even more, an object that has attached to it a larger surplus of circulating abstract desire that is in excess of the functionalist need for the specific object. This surplus of desire is not beside the point but is, in fact, like surplus value, crucial to the very success of the commodity form.[28] The commodity is wanted because there is want. In a similar way, GDP and other phantasmagrams make sense because of the surplus of affect that makes quantification lively, a surplus of feeling and immaterial presences that were not peripheral to "the economy" but rather looped in, integral to its very workings, crucial to its ability to gather up the all.

One of Karl Marx's most important methodological contributions in his critique of classical political economy was the way he saw the commodity not as a reified thing but as an abstract ontology brought into being by social relations. It took waged labor, property law, and their assembly with machines and materials to make a substance into a commodity. In looking at a commodity, Marx demanded that one notice all the dead labor, all the past life-force alienated from people and sold as waged labor toward the production of things that they do not own, toward profits for other's pockets. Commodities are haunted by this dead labor, this alienated life, that is obscured when one is in a store purchasing a commodity decorated in its bright packaging that quickly turns to waste. Similarly, we can refuse the fetish of an equation as the answer to the question of how GDP is calculated. What unseen elements have gone into the making of GDP? What deadness or liveliness make it possible? What haunts GDP?

From its inception, GDP was designed to count some things and not others. For instance, GDP counts the final production of market economic activity (activities bought and sold with money) and the spending of the state. Thus, GDP makes a cut between what counts as part of the economy and what is considered extraeconomic. But GDP is not the first such cut. The long-standing question of what counts as taxable activity already has states sorting what counts as economic as distinct from other forms of life activity, such as cultural life or family life. Historian Ritu Birla places the production of the economic as a domain distinct from "culture" as a consequence of nineteenth-century struggles between British colonial juridical efforts to "free" the circulation of capital and vernacular Indian merchants who maneuvered to remove arenas of their activities from colonial reach, even if involving wealth or money.[29] Carole Pateman, in turn, shows how the very invention of liberal politics in eighteenth-century France, with its announcement of the right to individually contract, nonetheless carved an exceptional juridical space for patriarchal rule within the family, in which children and women were ruled by the father in an extraliberal, extramarket domain.[30] Charles Mills shows how race logics, manifest in nineteenth-century European colonialism and American slavery, carved zones of exception from the liberal political versions of economic man through racial dehumanizing. Within the founding of liberal political frameworks that persevered from the eighteenth century into the early twentieth century, designations of less-than-human denied some the capacity to contract for themselves and own property. Even more perniciously, the racial logics of liberal political frameworks authorized the enslavement of people as property and the legality of stealing from, kidnapping, injuring, raping, and killing racialized less-than-human life. Economic life was thus constituted through the violent extractions of extraeconomic life. In other words, GDP has many ghosts.

Hence, the national income accounts and GDP are only a more recent slice in this longer history of constituting the economic by delineating the extraeconomic. Moreover, GDP counts the income produced from waged labor, selling commodities and services, rent, state spending, and financial practices of speculation and investment. It does not include unwaged labor, including unwaged labor in the home or for subsistence. As an oft-quoted American economics textbook explained, if a man married his maid, the GDP would fall.[31] In other words, gendered unwaged labor within a nuclear family does not count. Since 1998, income from speculative financial services, with no tether to the physical making of anything, raises GDP.[32] In calculating financial income, GDP does not care about levels of debt or degrees of damage; it only measures the returns. Therefore, GDP includes all the paid activities involved in responding to a disaster. The recovery efforts after a hurricane that devastates a city, the repair of a war-ravaged town, and the cleanup after a chemical spill all generate data that feed GDP. The loss of life and world produces no subtraction. Thus, GDP includes all the bullets, warheads, tanks, tear gas, and drones of war as an addition. It counts the war machine or police state as wealth but does not reckon loss and destruction. It does not include trading tomatoes from your garden for your neighbor's labor plowing a field, or any other act of barter. It does not include growing and preserving food for yourself; building and caring for your home; the work of seeking employment; the labor of caring for children, the sick, or aged; nor the wealth created by growing plants and tending chickens, goats, waters, or ecologies if they are not sold and you are not paid. Or even if your activities involve money but are illegal or informal, and hence viewed as incidental, they will not count; no data will be generated. All these activities are not part of "all the economy." Thus, the life-sustaining activities of people and the often gendered unwaged work of tending to bodies and communities do not count. They do not register in the calculus but haunt it as zones of life that could yet be subsumed.

In a map of the world with color-coded countries keyed to GDP per capita (figure 1.2), the eye is invited to view the globe as a patchwork of economic difference. It is a potent and familiar phantasmagram. By the end of the twentieth century, GDP sorts the world into have and have-not nations. Following midcentury decolonization, the landmasses of the planet are carved into nation-states each with their own macroeconomy to be developed, planned, and grown.[33] The practices for keeping national income accounts and tabulating GDP became globalized in 1947 through the de-

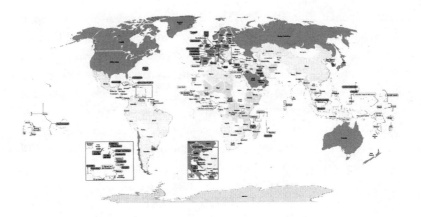

1.2 A typical World Bank map of the world (2012) carved into nation-states that each contain a macroeconomy. The eye is able to take in the world divided into territories of GDP, and GDP inequities, in one glance. The World Bank uses IMF data, and then its own proprietary way of converting GDP into a measure they call GNI per capita. (World Bank, World Developmental Indicator 2012)

velopment of United Nations (UN) standards that are repeatedly refined: 1953, 1968, 1993, 2008. British social scientists, operating through colonial networks, were crucial to this globalizing of GDP. In one of the first such endeavors in 1940, economist Colin Clark, who was pivotal to setting up the British national accounts, organized a monumental survey of the economies of all the states in the world, chasing estimates necessarily based on slim data, as only a handful of the richest industrial nations had built the kind of statistical apparatuses necessary for a regularized GDP tabulating practice.[34] He concludes that most of the world should be categorized as wretchedly poor, but also that national accounting is inadequate to this task.[35] Should separate accounts be collected for "modern" and "traditional" economic activities? How does one capture the wealth of the nonmonetized work of living, surviving, and caring for others that makes up the bulk of human activity? When the discipline of development economics, a field itself only invented in the 1940s, takes as its premise the goal of expanding "modern" economic activities, then GDP and national income accounts become adequate to this singular goal, as GDP will track only the hoped-for expansion of waged labor, factories, and industrialized agriculture that was the mid-century dream of economic development.[36] It will not track the debt and the damage, nor the sustaining and caring, nor the distributions. The GDP will track how countries "develop" in this narrowest of economic senses, and by the 1960s, the International Monetary Fund (IMF) will tie its loans to the ability to model economic development and grow GDP.

By the end of the twentieth century, if a country does not calculate GDP, the IMF will estimate it for you. And so GDP becomes a tyrant: a singular ruling index that demands the rearrangement of the world, of lives, of homes, of lands, of property, and of relations. The GDP demands to be fed. It must grow. It eats labor and rearranges life, asks for open borders for investment and commodity flows, demands the privatization of services and resources, the industrialization of agriculture, the "structural adjustment" of economies. The GDP is aspirational and desires to fatten. It gathers the numbers that dollars make and hides behind a single tally that binds all nations together in comparison. What has not been done for GDP?

As with any tyrant, the rule of GDP is widely denounced. It can grow even when there is mass poverty, as GDP does not care about inequalities in distributions of wealth, only its final accumulation. Hence, a whole "dashboard" of indices is needed to expand what is known about the macroeconomy, as many economists have come to argue.[37] Through the dashboard metaphor, visualized as a cockpit or instrument panel, the economy becomes a phenomenon driven by experts who adjust the indices that performative measure or predict the economy. Indices are the levers that help planners drive the economy, so more measures offer more ways of intervening. Or better yet, a differently designed index might better get at well-being and serve as a superior indicator of "development." An alternative index was created at the United Nations once the Cold War ended: the Human Development Index (HDI).[38] Led by Pakistani economist Mahbub Ul Haq and Noble Prize–winning Indian economist Amartya Sen, HDI was produced as a measure that equally remixed life expectancy and education with GDP in its tabulation. The HDI demands data on literacy, years of schooling, and vital statistics. Drawing on the "human capabilities" approach of Sen, the HDI urged investment into what Ul Haq called the social and not just economic "human balance sheet" of nations.[39] The HDI is joined by the Gender Inequality Index (GII), which measures maternal mortality, teen-

Janet Yellen's Dashboard

As chief pilot of the Federal Reserve, Chair Janet Yellen watches an array of gauges to steer the economy toward price stability and maximum sustainable employment. We've collected the most important to create Janet Yellen's Dashboard.

Inflation Growth Markets Jobs

1.3 In this 2014 illustration commissioned by the United States' Brookings Institution think tank, Janet Yellen, the powerful chair of the U.S. Federal Reserve since 2013, is depicted as steering a dashboard of her chosen economic performance indicators. Online readers are invited to click on the various dashboard dials and see "what the Fed Chair sees as she makes predictions for unemployment, inflation, growth, and interest rates." (Brooking Institution, Janet Yellen's Dashboard, 2014)

age pregnancy, numbers of women holding political office, education levels, and participation in waged labor. It also joins the Multidimensional Poverty Index (MPI), which "complements monetary measures" by consolidating thirty-three different indices of poverty.[40] Development measures become their own extrastate industry with cadres of social scientists gathering data, running sample surveys, crunching numbers, executing experiments, and feeding indicators. Poor countries become data rich.

In 2000 the UN crafted a new dashboard for development, the Millennium Development Goals, with over sixty indices: employment to population ratio, proportion of population below minimum level of dietary energy consumption, contraceptive prevalence rate, proportion of population using an improved drinking water source, and the proportion of urban population living in slums is all summed.[41] A global data infrastructure recursively grows and the numerical augury multiplies. The atmosphere of the macroeconomy has become denser and denser with indexable relations. There is no need to be a Keynesian about fiscal policy to believe in the macroeconomy.

Decolonization offered other possibilities of envisioning how a nation-state might care about the economy, often refashioning the Marxian tool-box for critiquing political economy in attempts to craft another way. Because GDP does not care about ownership, and hides the foreign extraction of wealth and life from the veins of previous colonies, it grows even as it dispossesses. In the 1970s, Mariarosa Dalla Costa and Selma James, among others, argued that decolonization needed to simultaneously address colonialism's racism, persistent patriarchy, and capitalism's sexual division of labor, in which women lived in poverty doing the unwaged work of sustaining human life by caring for children, feeding families, and practicing subsistence agriculture.[42] Capitalism, they argued, not only devalued this labor but relied on it as the free work that reproduces the labor force. Women were both a reserve army of labor that capitalism could call upon or dump as need be and a source of free labor that made laborers. The economy relied on women's embodiment itself—their breasts, wombs, and feelings, their fleshy vitality—to create future workers. Marxist feminists of the 1970s named reproduction as a kind of *unwaged* work, coining the very term *unwaged*.[43] Thus, they organized campaigns, such as Wages for Housework or the Global Strike of Women, that attempted to revalue caring work as a means to rearrange the values of the nation-state that counted killing over caring.[44] As an epistemological effect of this political work, Marxist feminists helped to carve out "reproduction" as a politicizable economic domain. Reproduction happened in the shadow space of economy and needed to be brought to light.

What counts as "reproduction," what is included within its ambit, is as much a question as what counts as "production" in the "economy." As concepts, production and reproduction have been profoundly entangled in their respective genealogies and are at stake in one another.[45] At its broadest, reproduction refers to the "action or process of forming, creating, or bringing into existence again," according to the *Oxford English Dictionary*. The term *reproduction* was used by eighteenth-century economists such as François Quesnay and Adam Smith to describe the process of re-creating economic value: money invested in the manufacture of a commodity was reproduced once it was sold. In the seventeenth century, reproduction had referred to living processes of coming into being again such as regenerating or remaking tissues or limbs, as in the example of a salamander regrowing a leg. At the end of the eighteenth century, when the French Revolution exploded a new liberal political order offering legal and ethical form to capitalism, the famous French comparative anatomist Georges-Louis Leclerc, comte de Buffon, was the first to use the word *reproduction* (instead of *generation*) to name the process of organisms coming into being. *Reproduction*, through Buffon, came to name a process of maintaining a species in time, a process that perpetuates the stability of form in organisms across generations.[46] Thus, reproduction as attached to life and birth was initially conceptualized as an extensive process, binding and producing an aggregate living formation — the species — working through and creating the individual organisms of that species. In other words, eighteenth-century *reproduction* was a process of replacement, sameness, and consistency that linked generations of embodied individuals together as a persistent common kind over time. Importantly, *reproduction* as an eighteenth-century term moved between natural history and political economy. In either domain, reproduction worked through laboring bodies (in work or birth) but exceeded them.

Within the evolutionary epistemologies of the nineteenth century, *reproduction* continued as a species-level process of living-being, yet one that occurred in extensive spans of time, constituting the ongoing historicity of life.[47] Moreover, reproduction as an evolutionary process did not simply maintain species-kind but instead produced the difference that natural selection sorted. Reproduction as part of natural selection did not just remake life the same; it produced a becoming-in-time that generated populations composed of variation, of which only some survived. With evolution reproduction had become a living difference engine. Reproduction stretched beyond mere individualized sex, to occur at time scales larger than the life of the organism, into the recesses of evolutionary time and hence species change. By the early twentieth century, a technocratic ethos was applied to this production of population-level variation. The future of reproduction could be dreamed through its rational guidance by technoscience, giving rise to the genocidal project of eugenics and the pernicious logics of racial difference.

In the nineteenth century, Marx too theorized *reproduction* at the cusp of economic and living-being. In *Capital*, volume 1, *reproduction* named the return of some of the products of accumulation back into the means of production: the machines, material, and labor that assemble together to produce. For capitalist society to maintain itself over time, it needed to keep replacing the means of production. This reproduction happened in the business of a factory but also in the life of the worker, who must spend wages in order to pay for his own subsistence and maintain his life. Laborers must reproduce themselves from day to day. Reproduction, for Marx, was also the name for the process by which life and labor power continued themselves in time, both in terms of eating and sustaining the life of an individual body and in terms of aggregate life, the life of workers across generations: "If production be capitalistic in form, so too will be reproduction."[48] As a system of relations that assembled together machines and bodies, capitalism was continuously remaking those relations through "a process of reproduction, [that] produces not only commodities, not only surplus-value, but it also produces and reproduces the capitalist relation; on the one side the capitalist, on the other the wage laborer."[49] Marx notes, but does not elaborate, that the activities of women and children in contributing to the subsistence of the family are a form of "supplementary labor" that produces the "labor power" of the worker for capitalism.[50]

Marxist feminists in the 1970s amplified this element in Marx as the unwaged "reproductive labor" of women in maintaining life within economy. Inhabiting the negative space produced by the GDP and national income accounts, Marxist feminists drew into legibility a domain of reproduction that attached sexed living-being to the macroeconomy as its thickly inhabited shadow space. Capitalism relied on reproduction. Yet just as feminists were bringing into new visibility reproduction as the devalued labor of maintaining and creating life, reproduction as construed through the figure of population was also coming into legibility in the 1970s for economists and social scientists as crucial, rather than incidental, to the success of the macroeconomy. Reproduction, through the social science figure of population, would become tightly tied to the very fate of the macroeconomy.

While initially the measurement of GDP placed reproduction in the zone of extraeconomic life, in contrast Cold War demography would plot the historical progress of the economy in a way that firmly tied it to reproduction as a feature of population. Cold War/postcolonial demography's own quantitative models would plot rises and falls of mass births as vital to the economy's future, particularly for recently decolonized countries, inciting a new round and another layer of epistemic infrastructure built toward counting and governing economy and population together. It would be this Cold War/postcolonial conjugation of population and economy that would propagate new techniques for valuing—and devaluing—human life.

In 1959 Frank Notestein, an eminent American demographer, stood at a lectern at the newly founded Institute of Development Economics in Karachi, Pakistan. Notestein's concern — "Abundant Life" — hinged on the play between abundance as quality and quantity.[1] He argued that future economic growth, and hence "the good life" of modern production and consumption, is dependent on the reduction of population growth. In short, reduction in future population quantity leads to good economic quality.

Also in 1959, Eisenhower's presidential committee on military assistance, chaired by General William Draper, concluded that supplying arms and military training to strategic "front line" nations, though essential, was not sufficient to ward off "imperial" communism and thwart the "Soviet Economic Offensive."[2] Draper's report unexpectedly called for economic assistance combined with population control, expanding the Cold War into realms of reproduction. For Draper, swelling poor populations in "less-developed countries" threatened to undermine the expansion of capitalism, hence birth control became a national security solution.[3] It is this report that recommended the founding of a single agency to administer foreign aid, giving birth to the United States Agency for International Development (USAID) as the international economic development arm of the U.S. state. If poverty was a breeding ground for Communism, then only by funding population control in decolonized frontline sites could the United States, according to Draper, extract "the maximum result out of our [military] expenditure" and achieve mutual security against Communism.[4] Birth control, in its military function, would work to stem the tide of Communism as well as offer a before-life/pre-death delivery device — contraception. Birth control would prevent the need for more conventional weaponry, and hence future deaths. In this way, contraception became a preemptive strike against both the purported population explosion and future war.[5] Population pressure, so social scientists such as Keynes argued, had been a recent cause of war in both Europe and the Pacific, and thus war might be prevented by reducing future births. The problem of population was represented as the Cold War "population bomb" in which the birth of others threatened to detonate not only the American way of life but the planet.[6] From this point onward, a Cold War function for family planning underlay much of the tremendous flow of funds from the United States in the 1960s and 1970s, not only into official state family planning projects but importantly into a new organizational form, the NGO and private nonprofits, which strategically kept family planning services at arm's length from states and the Cold War strategy that helped fund them.[7]

Supporters of the Draper report could marshal evidence for their case from a Cold War investment in social science and demographic research that measured "levels of living," "cost of children," and again "rates of increase." Notestein was among the most significant U.S. demographers in this effort to craft governable population models and measures. His signal contribution concerned the "demographic transition," a promissory and universalizing model that pegged temporal changes in fertility and death rates to economic development.[8] This model, based on Europe's past, predicted forward to the future of recently decolonized countries entering so-called modernization.[9] The demographic transition model (still widely used in policy, even though its veracity is heatedly contested within academia) translated the past of the European metropole onto the future of previous colonies.[10] It also revised Pearl's S-curve of population by more explicitly disentangling birth and death as their own lines to be plotted onto stages of human historical time. While Pearl modeled how human population growth changed relative to the economic environment, the demographic transition model tightened the coupling of population and economy. Economy did not just set the conditions for population; the reverse was also the case: population dynamics could promote or undermine economic growth. These two aggregate forms of life were modeled as mutually setting the terms for one another.

The demographic transition model, as expressed through Cold War modernization theory, divides the axis of time into stages of economic "development."[11] Stage 1 is correlated with a "traditional," "premodern," or "agrarian" society, in which fertility and death rates are both high, and thus population growth is flat. In stage 2, the introduction of modernization (via colonial public health, for example) is correlated to a reduction in death

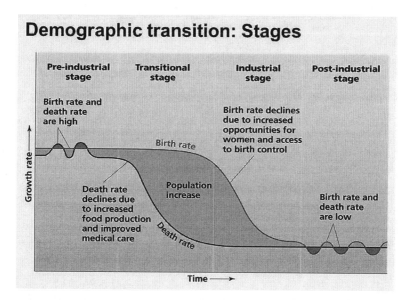

Demographic transition: Stages

Pre-industrial stage · **Transitional stage** · **Industrial stage** · **Post-industrial stage**

Growth rate

Birth rate and death rate are high

Birth rate

Birth rate declines due to increased opportunities for women and access to birth control

Death rate declines due to increased food production and improved medical care

Population increase

Death rate

Birth rate and death rate are low

Time →

2.1 A demographic transition model commonly found in textbooks in sociology, demography, geography, and biology. This particular illustration from 2006 has been repeatedly used by the Pearson line of textbooks as well as their prepackaged lecture slides across multiple disciplines. It portrays a standard and idealized narrative of the demographic transition model, dividing it into four stages of human economic development, from preindustrial, to transition (the term used here for the midcentury period of "developing" during colonization and decolonization), and then on to the ideals of industrial and postindustrial phases (the time-space of the West). Plotting time against population growth, the model narrates a prior stage of high birth and death rates (and thus stable population growth), which gives way to a transitional, or developing, stage where modern interventions have decreased death rates, but not birth rates, leading to rapid population growth. Based on a purported trajectory for Europe, an industrial and fully modern period will feature low birth rates, decreasing population growth, toward an idealized postindustrial horizon of stable population. This universalized model leaves open the question of whether lower birth rates help to usher in modern capitalism, or whether the arrival of industrialization stimulates a desire for less children. While the chart narrates a linear and universal progression, it does not give the timing of the different phases, which are understood to vary from place to place, and hence these differences in timing underwrite the need to manage and quicken the transition to the next phase in particular places that are slow to progress. Note there are no data used to derive this graph; it is instead an abstracted and highly simplified model of purported relationships that directs imaginations about the history and future of population. (Pearson Prentice Hall, Demographic Transition Model, 2006) rates, but not birth rates, so that population growth takes off and threatens to undo the thrust of economic modernization by outstripping economic production. In stage 3, modernization in the form of industrialization, consumption, changing attitudes, urbanization, and state investment in public health causes fertility levels to decline and population growth to subside. In stage 4, full modernization is reached and fertility and death rates are now both low, leading again to a stable population.[12] Thus, the demographic transition model offered a reframing of colonial habits of carving the world into stages of primitive and modern, forward and backward, but in economic terms. It offered a universalized vision of industrial capitalism manifesting in every part of the world for the sake of future aggregate life.

Moreover, the demographic transition model framed "the long view" of population.[13] Like Pearl's S-curve, individuals disappeared at the resolution of multigenerational aggregates. So too did proponents of the demographic transition model assemble a globalizing view, marshaling population forecasts from every corner of the globe, with charts on New Zealand, Oceania, Korea, or Japan echoing findings from France, Egypt, and Mexico. The demographic transition could happen everywhere, but at different paces. Moreover, it was a model based on broad correlations between population and economics, leaving precise causes open to debate: was the decrease in birth caused by social conditions, technological changes, or industrialization? Or was fertility decrease itself a prerequisite of economic development? Finally, the demographic transition model hinged economic futures on questions of sex. High fertility became the variable in need of adjustment for the sake of economy. Economy pivoted on sex.

For Notestein, the demographic transition model was more than abstraction; it guided his practical policy work. Notestein was less an academic than a new kind of postcolonial expert.[14] He was a charter member and third president of the Population Council, overseeing its international expansion of family planning projects in the 1960s. He was the first director of the United Nations Population Division (opened in 1946), as well as the founding director of the first U.S. university program on population, the Office of Population Research at Princeton (1936). Notestein was a formidable force guiding the research direction of a new generation of population experts, many of whom he directly mentored. He developed his theory of the demographic transition largely in speeches instead of academic papers. In practice, the demographic transition became the single most important model through which national population policies around the world were crafted.

Standing at the podium in 1959, then, Notestein was a Cold War and postcolonial policy consultant performing a ritualized expertise invited by Pakistani state planners working under the military dictatorship of Ayub Khan, a so-called military modernizer. Ayub Khan had risen to power through a coup to become a strategic U.S. ally and recipient of military aid. Already by 1959, Ayub Khan was convinced of the need for population control, actively seeking out collaboration with Notestein and the Population Council, while publicly affirming the seriousness of the "menace of overpopulation."[15] The Pakistani statisticians and economists who presented alongside Notestein were at that moment crafting their second Five Year Plan, which like India's state planning model required the intensive generation, analysis, and modeling of quantitative data. After partition, Pakistan had to gather a patchwork of regional governance structures into a new national bureaucracy. In contrast, India had inherited more national and colonial state infrastructure, and quickly established a National Income System as well as statistical and demographic institutes to track the quality and quantity of economy and population as planning objects. Prasanta Chandra Mahalanobis, the architect of India's state planning model and a famous statistician, was trained in the Galton Laboratory at King's College, and like Pearl started off with eugenically motivated research that, by midcentury, was superseded by commitments to statistical rigor.[16] He is credited with developing a unique planning model for India, crafting a "third way" between American-style capitalism and Soviet-style planned economies. In turn, the Indian economist V. K. R. V. Rao, who received his doctorate at Cambridge under Keynes, established India's national accounts, which he had tabulated as early as 1934, simultaneous to the United States and the United Kingdom.[17] With independence, South Asia had become an exuberant node of postcolonial social science and as such was a crucial site for elaborating new governmentalities for guiding population and economy together.

After partition, Pakistan had to invent a new national state infrastructure distinct from India, and severed into two noncontiguous geographic areas separated by more than one thousand kilometers: Urdu-language majority West Pakistan and Bengali-language majority East Pakistan. The state apparatus of Pakistan was governed by a small corps of elite civil servants (not famous intellectuals) ensconced in West Pakistan, who in the years immediately following partition were still trained in the British colonial style of administration.[18] As a result, West Pakistan ruled over geographically separated East Pakistan. Statistical analyses, social science modeling, and civil service hierarchy enabled rule from a distance. A small bureaucratic elite with often technocratic aspirations developed top-down planning goals that were ordered to be fulfilled by local administrators equipped with standardized forms and procedures, themselves unevenly and arbitrarily distributed between provinces as well as between rural and urban spaces. This particular modernist and technocratic top-down approach to governance by paper and file was reflected in the building of Islamabad as a planned capital under Ayub Khan that served to both isolate the civil service elite and rationalize the physical landscapes of governance.[19] Despite technocratic modernizing aspirations, in practice Pakistani planners had to contend with the paucity of quantitative data about both demography and national accounts.

The particular cohort of Pakistani social scientists who met with Notestein, as Pakistani left critics at the time noted, was part of an elite circuit of training and advising with the Harvard Advisory Group,[20] which was intimately involved in developing Pakistan's second Five Year Plan. Harvard economist Gustav Papanek was particularly influential in crafting a policy of "functionalist inequality," which held that increased inequality, and hence cheap labor, could serve to raise GDP and industrialism, and thus economic policies should shelve questions of income equality for the future.[21] Some members of the Harvard Advisory Group explicitly considered Pakistan their "laboratory" for their economic development theories.[22] Within a Cold War circuit, a thick traffic of U.S. social scientists worked with Pakistani planners as an important experimental field site of the decolonized world. Reflecting back on his work in Pakistan, one Harvard Advisory Group member observed the rise of a formation of experts: "Since the last war, a new type of human being has come into existence. He is the foreign advisor, the expert, the specialist—call him what you will—who has replaced the colonial service officer."[23] Both Pakistani left critics and U.S. practitioners remarked on the emergence of this new "rule of experts."[24] U.S. and South Asian social scientists were swept up in technocratic visions of new modernities even if differentially imagined by anticolonialists, postcolonial economists, military dictatorships, U.S. foundations, and U.S. foreign aid. In turn, communities, and one might say the entire citizenry of Pakistan, were hailed as potential experimental subjects in a postcolonial laboratory.[25] Each instance of economic development policy at this moment was a trial gathering the majority of the world into policy experiments. Within this efflorescence of experts and experiments, the aggregate of "population" was continually, reiteratively reassembled and not merely studied.

Notestein's specific enactment of expertise, presented at the Karachi seminar, was thus a staged polemic for the need to co-plan aggregate fertility and economy. The Pakistani economists present would themselves become significant global neoliberal experts. The event was performed with the oversight of the finance minister, economist Muhammad Shoaib, who would later serve as a vice president of the World Bank. The seminar itself was published by economist Moeen Qureshi, who would later work for the IMF and then serve as the interim prime minister who ushered in structural adjustment to Pakistan.

On cue in 1959, Notestein declared Pakistan at a particular stage of modernization: "in the immediate future . . . there is a conflict between qualitative and quantitative abundance of life."[26] The purported compressed pace of modernization—in Pakistan now compared to the norm of Europe's past—called for the purposeful reduction of population growth to support increased economic productivity (typically measured in GNP per capita) and thereby avoid projected catastrophe. Here, the fertility rate (as potential future lives) needed proper adjustment—with contraception and sterilization—to manage the temporal shift into modernity. Notestein's "abundant life" was symptomatic of an *economized* reformulation of Foucault's description of the violent purifications of state racism as some must die so that others might live into *some must not be born so that future others might live more abundantly (consumptively).*[27]

"Abundant life" articulated a new calibration of value connected to the internationally elevated metric "GDP per capita," the globalized index for comparing the economic vitality between countries of different size that was bluntly determined by dividing total GDP by population. In the demographic transition model, GDP per capita became the quantitative measure of lives less "abundant" and lives "underdeveloped." In 1958, under the auspices of Notestein and funded partially by the World Bank, demographer Ansley J. Coale and economist Edgar M. Hoover collaborated on a new model of the demographic transition based primarily on data from India (but applied universally) that amplified and thickened the effect population had on GDP. Celebrated as one of the most influential models in modern demography, Coale and Hoover's work took as its premise that as families had more children, their savings would decline (called "propensity to save").[28] Further, society would have to spend more money to educate and care for these children, even though many would not survive to adulthood, creating a reduction in the efficiency of the state's overall investments in the population (called its "incremental capital output ratio"). Investing in children who "because they die make no contribution at a later time to the economy" is a kind of economic "waste." Yet even a reduction in mortality does not solve this problem, as "the 'waste' of always supporting a much larger next generation (the waste of very rapid growth) replaces the waste of spending on persons who die later."[29] At their macroeconomic modeling level, these trends reduced GDP further, such that an increase in population had a cumulative *negative* effect on GDP beyond its effect as a denominator of GDP per capita. More lives were detrimental to GDP growth; they were negative in value. Ayub Khan's own position on population control was influenced by this model.

Coale and Hoover's work amplified how the demographic transition model allowed a differential valuation of human life to be explicitly monetized in ways foreclosed at other registers and scales. First, it offered a way to govern human population growth at massive scales and temporal distance—through intensive abstraction and the long view—such that calibrations of human worth were removed from consideration of any specific actual life. Feminist critics within demography would later point out in the 1990s that the demographic transition model was notable for the way it erased individual women, who were typically the real-world users of contraception, targets of population policies, and fodder for the data collection that made such models and policies possible.[30] Second, the dissolution of individual people within "aggregates of humans" was reinforced by the focus on rates rather than kinds of people: rates of growth, contraceptive prevalence rates, or birth rates. Expressing policy goals in terms of altering rates deflected from the implementation of population policies that targeted specific human lives already shaped by class, sex, race, and caste. Third, the demographic transition model extrapolated its temporal scale into the future, thereby prompting a speculative and preemptive logic that targeted *potential lives*, lives that did not yet exist, via interventions into current lives. Lives-yet-lived (babies not yet born) were the primary entities devalued and "averted." All these features worked together to make an explicit economized calculation of differential life worth palatable as an aspect of governance.

Notestein's own formulation of the demographic transition was tightly bound to a Keynesian commitment to state planning of the economy, with an emphasis on consumption, labor, and savings. However, broader than any Keynesian frame, the demographic transition model was an epistemological technology that circulated widely because it could support diverse population control and family planning policies. It could be mobilized within socialist planning models, as with the One Child policy of China, as well as in projects with emerging neoliberal logics, such as those fostered by USAID that responsibilized women to choose their own fertility by presenting them with a limited "cafeteria" of contraceptive options.[31] In other words, the demographic transition modeled an abstracted rate in need of adjustment, where the exact form of the intervention (voluntarist, racist, violently coercive, feminist, and so on) as well as the causal claims underpinning it were not a priori scripted.[32] The urgency of the demographic transition could authorize the coercive mass-sterilization camps of India's state of emergency in the 1970s and it could justify investment into women's work cooperatives in Bangladesh inflected with a feminist politics that sought to prompt family planning obliquely by enhancing women's independence. The demographic transition model did not provide a *law* of population (as Pearl claimed for the logistic curve) but rather an abstract anticipatory instrument that could mobilize an enormous range of interventions.

While the demographic transition model was promoted in transnational postcolonial circuits, within the Cold War United States a new throng of experts enthusiastically offered up "population" to the wider public as a potent specter to be feared. At least since Pearl's experimental work on the problem of overpopulation, and accompanying the rise of "the economy" as the container that supports life, a thickly inhabited genealogy of U.S. experts (from sociology, economics, and biology) have repeatedly pointed to the steep upward slope of the graph of population growth as an urgent threat to the future of the "good life."[33] During the Cold War the field of demography and population science propagated, as did organizations oriented toward the popular transmission of concern for the population problem: the Population Council (1952), the Population Crisis Committee (1965), the Association for Voluntary Sterilization (1965), USAID's Office of Population (1965), and Zero Population Growth (1968). Fear of dystopic overpopulation was prophesied in apocalyptic best-selling books depicting humans as crowds, bellies, and mouths, such as William and Paul Paddocks's *Famine—1975! America's Decision: Who Will Survive?* (1967) followed by Paul Ehrlich's even more wildly popular *The Population Bomb* (1971), which was replete with frightening future scenarios of nuclear war and mass famine. Historian Emily Merchant characterizes *The Population Bomb* as a work of "dystopian science fiction," as the bulk of its narrative rested on spinning out terrifying speculative scenarios. Panic about population was spread by paid advertisements in the *New York Times* linking overpopulation with crime, and by cover stories in popular magazines such as *Life*, *Time*, and *Newsweek*. The simple line graph with its urgent upward curve that forecast a future overrun by sheer human numbers proved to have tremendous affective force. According to the graph, future poverty, hunger, and war would be the global fallout of a population explosion of brown bodies. Even the more scientistic *The Limits to Growth* (1972) report of the Club of Rome would rely on the affective charge of the line graph as the output of a computer-generated world simulation that portended a global collapse caused by the dual upward trajectories of economic production and population. Explicitly incorporating the demographic transition model, *The Limits to Growth* simulation factored in an anticipated decline in birth rates that would accompany an anticipated increase in GDP, yet in their model the finite container of the earth would deplete over time, the bottle becoming smaller and smaller, so that overpopulation was almost inevitable. Inside the model—presented as a set of superimposed line graphs—were the same assumptions about the differential value of children relative to GNP, a "rough cost-benefit analysis."[34] While the demographic transition's speculative futures could herald

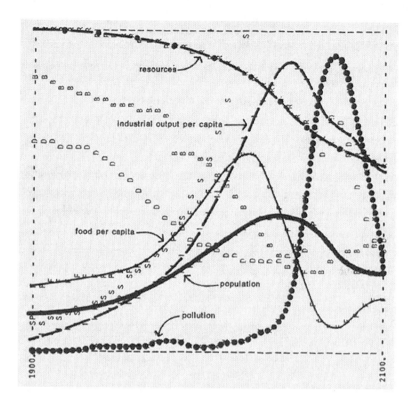

2.2 Graph from *The Limits to Growth* project of the Club of Rome. Runs of its computer model were displayed with line graphs that inevitably traced an urgent rise and fall for "mankind." In this run of the model, the variable of resources was set to unlimited, and "perfect birth control" was added to the estimates of population growth. But even here, the model predicted the crash of population, the falling line indicating the future mass death of humanity, much like the graph of Pearl's fruit flies. (Meadows et al., *The Limits to Growth*, 1972)

optimistic possibilities of abundant life in postcolonial sites, in the United States the demographic transition was decidedly animated by fear of a future with too many of "them" that would derail the American good life of capitalism and white supremacy.

With the demographic transition as the phantasmic guide to the postcolonial future, the speculative interventions promoted transnationally were necessarily experimental: intervene, gather data, recalculate, revise. Numbers gathered under the logics of the demographic transition abstractly calculated the value of possible future lives (and the prevention of such lives) as tied to economic time. Lives not worth being born abstractly held a negative economic value that would be attached to the possible life of future people, particularly nonwhite people and poor people. Out of these quantitative models a new way of distributing life chances and the value of life had been built that did not need to rely on biological race but could continue to reenact racism. Some must not be born so that future others might prosper.

Averted birth is a figure of the better-not-born, a naming and counting of a better-to-have-never-lived. Emerging in the 1960s in the United States, it was a new calculative figure of devalued or "wasteful" life to be prevented. Averted birth was often an anticipatory measure. It counted toward a future in which particular lives do not exist. Perhaps the most influential of such calculations was performed by U.S. economist Stephen Enke, who demonstrated that money spent for each "averted birth" was "100 times more effective" in raising GDP per capita than the same amount spent on "productive investments," a claim that helped to further spawn cost–benefit analyses for specific family planning interventions.[1] Enke's work was crucial in convincing President Lyndon B. Johnson to order earmarked foreign aid funds for family planning over health, food, or kinds of aid, as well as making family planning a funded component of the domestic "war on poverty."[2] Enke's work influenced U.S. foreign policy on family planning throughout the 1970s, as well as the practice of calculating family planning targets in U.S.-based programs.

Caught in cost-effectiveness calculations, counting averted births was repeatedly posed as a question of dollars. How much does it cost for a program to deliver an averted birth? Which contraceptive was most cost-effective for creating averted birth? What is the value of a life not born? This is precisely what Enke's work sought to figure. Enke argued that "in almost all poor and backward countries" children's worth is negative for the national economy. Given the positive effects to GDP caused by reduced fertility, Enke calculated that "the 'worth' of preventing a birth in a typical L.D.C. [less developed country] is about 2–6 times the output per head."[3] In other words, averting a birth contributed more to GDP than the average labor of a living adult. *Unborn lives were worth more to "the economy" than lived lives*: "From the viewpoint of the national economy at large, additional children do have a plus or minus, monetary value. In point of fact, in almost all poor and backward countries their discounted 'worth' is today *negative*."[4]

Lives lived were "negative" in their value. The figure of averted birth did more than devalue future life; it cast a shadow over living people, who were also better-not-born.

Averted birth allowed national family planning projects to explicitly and quantitatively orient toward preventing life. Attempts to assign averted birth targets accompanied most national family planning projects by the late 1960s, and influentially were first reckoned for Taiwan's IUD (intra-uterine device) program.[5] Calculating averted births was intimately linked with cost–benefit analyses, such that planners might declare, "2.3 Million Births Averted in Korea, Savings Huge."[6] In the late 1960s and early 1970s, the question of how to calculate averted births was a mathematical challenge for demographers.[7] As a count of births that had not happened, the calculation of averted birth was another kind of phantasmagram: it conjured something profoundly immaterial, retrospectively tabulating events that had not occurred, or predicting forward to events that would not happen. One influential calculation of averted birth described it as a kind of "borrowing from the future" because of the ways its calculation depended on a host of other forecasts and probabilities.[8] To arrive at the number of averted births, one had to estimate the counterfactual number of births that would have occurred without any family planning intervention. This counterfactual calculation, in turn, relied on estimates of the average "foetal wastage" of miscarriages and anticipated infertility of women due to age, as well as estimates of the average time spent married, of time spent pregnant leading to a live birth, which was distinct from the also required estimate of time spent pregnant leading to a still birth, which was different again from the estimate of average time pregnant leading to a miscarriage. Moreover, each of these categories of pregnancy required the estimation of the time spent "anovulatory" after each of these outcomes, as well as after an abortion.[9] To calculate averted births from sterilization (which assumed a perfect success rate), actual births were subtracted from the counterfactual calculation of possible births. But for other contraceptives, such as IUDs, variables such as "retention rates" (the rate at which women did not quit a contraceptive) and estimates of imperfect contraceptive performance were also factored into the formula as probabilities. Moreover, in the 1960s and 1970s, the models for calculating averted birth were designed with hetero-normative and patriarchal premises of a world in which only women who were married have babies, refusing to place in its model the figure of someone who gets pregnant in other ways.

Projections of averted birth were thus arrived at only through concatenations of many other estimates and projections, each fueled with the hopes and fears of population planners. Averted birth was phantasmagramically brought into being through a thick matrix of speculative and counterfactual measures, each imbued with aspirations of technical efficacy that had material consequences for the organization of state projects, infrastructures, and funding.

Averted birth was more than just an estimate generated in a flurry of other estimates. Babies not born were its spectral referent. The haunting presence of the not-born as the goal of population control moved one demographer to rephrase an American poem ("Antiginosh" by Hugh Means) describing a ghostly visitation:

Yesterday from dark to morn,
Ten thousand babies were not born;
Like autumn leaves the non-born fall;
I wonder where we'll put them all?[10]

Instead of a ghost created by a life lost to death, averted birth named another kind of unliving that was an undead-never-alive. It named not a singular specter but the dense presence of aggregate devalued life. Averted birth was both the elusive calculation of family planning programs and also an affectively charged presence in the world of could-have-been babies who could have been wanted or unwanted, whose absence could have caused reprieve or mourning.

While experts involved in calculating averted birth saw it as having the "abstract purity of a mathematical symbol," in practice averted birth manifested in women's lives as abortions, or in the consequences of sterilization (both voluntary and coerced) over a lifetime, or in the taking of daily pills that tied women into pharma circuits, or in the insertion of devices into intimate flesh that could be accompanied by side effects for which no medical infrastructure was there to care.[11] Averted birth could manifest as a forced tubal ligation performed in the intense and dizzying moment of delivering a child, or in the unknowing use of an experimental technology being tested at a family planning field site, or as the consequence of a thoughtful deliberation. Averted birth could appear as painful absence after the violent loss of fertility through a nonconsensual medical act. It could be an unnoticed absence from a decision taken long ago. As a calculus, averted births included all these possible expressions yet was indifferent to these lived differences. Moreover, models of averted birth were indifferent to women's intentions, as the measure does not distinguish force from choice. Calculations of averted births were made for national family planning programs around the globe but also for local programs in the United States where such mathematical ventures were undertaken to determine averted births among blacks in New York City or among poor teenagers.[12] As a count of aggregate unlife and a designation of devalued life, women did not manifest as persons but only as raced, classed, and aged estimates toward mathematical symbolization of a necropolitical specter.

Enke's work did more than assign dollar values to lives averted. His simulation work built computerized phantasmagrams of population that promised to imbue planners with a greater sense of how population could be intervened in for the sake of the economy. Drawing on techniques he had developed in the logistics department at RAND and for the U.S. Air Force, Enke helped to build population control computer simulations in which planners could be trained in enacting the production of averted birth, raise GDP per capita, and thwart the population bomb.[13] Enke called his hypothetical simulation country Developa. Simulations modeled the system of forces that might shape population dynamics and also served as a "teaching model" that could imbue planners with a felt imaginary of dangerous possible futures and the need for intervention. Built at General Electric's Technical Military Planning Operation (TEMPO) think tank and funded by USAID, the demographic computer simulations offered a pedagogy of feeling: simulations would implant "awareness" in national planners and "sell the demographic-economic approach" around the world.[14] The Carolina Population Center, also funded by USAID, embraced this pedagogy of simulation, building an interactive computer game to teach family planning administrators in "developing countries" how "to visualize and to choose wisely." Made in 1970 and used in South Korea and Taiwan, the Family Planning Administrator Training Game was among the earliest interactive computer games (preceding the famous Colossal Cave Adventure and Zork) in a moment before computers had screens but instead relied on printouts.[15] Gamers were promised the chance to manipulate a set of ideal-ized interrelationships that had been simulated as a hypothetical country. The success of their decisions would print out as a count of "averted births" and "cost per averted births" for the next round of play. The game was de-signed to teach players a palpable imaginary of population and economic dynamics, without the players having to understand the math behind the simulation, thereby encouraging a confidence that the policies promoted by USAID would result in the right outcomes. Through simulation, admin-istrators would hopefully become adept at countering the "irrational use of human reproductive powers" and produce cost-effective counts of the un-born.[16] Simulations gave users a pedagogy of feeling population and econ-omy together.

Planners would turn to such models and simulations to answer the prag-matic question of how to set population targets necessary to meet their GDP goals. Such population planning involved a cascade of wishful calcu-lation. Producing the metric of averted birth, the demographic transition model is among the most potent phantasmagrams in the economization of life. It persisted, and still persists, as a planning model even though most experts agree that it is empirically inaccurate. Today the demographic tran-sition gains influence with the addition of a fifth stage, where "advanced" economies (such as Germany, Taiwan, or Japan) are correlated with fertility so low that population declines, and thus threatens the national economy anew as an aging population that does not replace itself. Yet even the ques-tion of whether family planning programs have had much effect on fertility rates, or if fertility changes for other reasons, remains an open question in demography. The success of the demographic transition model was not in its empirical veracity but in the way it gave form to a technocratic dream of a national macroeconomy that could be fostered, directed, and triggered by rearranging reproduction en masse.

At midcentury, technoscience was dreaming postcolonial futures of many kinds: new worlds of rationalized equality, worlds without racism, worlds of industrial prosperity. The demographic transition was one such phantasmagram among many. Yet it called entire populations into its force field. It was a dream intensely felt by planners, with tremendous worlding consequences. It cast precarious life as a kind of surplus, a devalued and un-wanted excess amenable to erasure and optimization. It shaped the national policies of India and Taiwan, of South Korea and Pakistan, of Egypt and Thailand, of Kenya and Tanzania, of Jamaica and Puerto Rico, of Palestine and Bangladesh, of Haiti and Indonesia, of Japan and China, of Canada and the United States, such that it in some way touched nearly every place on earth. The demographic transition model mobilized the distribution of freely delivered contraceptive commodities as a core task of late twentieth-century American empire, and thus gave impetus to a pervasive presence of American-funded family planning commodities on the ground not just in South Asia but across East Asia, Africa, the Caribbean, and Latin America. Population was a Cold War threat with a planetary imaginary. "Over a billion births will have to be prevented during the next 30 years to bring down the world's population growth rate. . . . The task may well be the most difficult mankind has ever faced, for it involves the most fundamental characteristic of all life—the need to reproduce itself."[17] The now declassified 1974 "Kissinger Report" titled *Implications of Worldwide Population Growth for U.S. Security and Overseas Interests* characterized population in developing countries as threatening "severe damage to world economic, political, and ecological systems." In this risk assessment, overpopulation not only exacerbated underdevelopment but could become a "volatile, violent force."[18]

Population as modeled through the demographic transition justified the enormous flow of funds given to family planning through USAID and then passed onto NGOs (funds that could then be variously appropriated and re-directed to build small feminist and community projects of many kinds). The dream of the demographic transition built an enormous, globalized, variegated infrastructure that conjugated family planning services with economic planning. It built measures, and indices, and counts. It gave anticipatory form to the future of a postcolonial globe by reproducing yet another form of global social science rule.

While violence is integral to the phantasma of the demographic transition model, this does not imply that the phantasies technoscience generates are therefore false or made up. The phantasmagram does not undo objectivity; it instead adds the trust and belief that animates numbers, or the fear, anxiety, paranoia, or hope that orients and motivates facticity. Phantasy adds the palpable sense of the large immaterial forces that models aspire to glimpse. In the busy world of simulations, stochastic models, and mathematical symbolization, phantasy was the lively surplus of feeling, awareness, worry, hope, and imagination that immaterial phenomena required to become present to planners and experts, to bureaucrats and feminists, to dictators and citizens. Quantitative modeling is more than the mere tabulating of numbers. It is also the infrastructure and the practices of knowledge-making that perform the count, the galaxy of interrelationships imagined beyond perception, the specters of the uncounted that are called into palpable relief, and the affectively charged aura these arouse. Quantitative models and measures produced at the conjuncture of economy and population were enlivened by a surplus of phantasma integral to their very workings and their spread across the globe. We might then think of phantasmagrams as having material-semiotic-affective-infrastructural ontologies.

Walter Benjamin, in writing about the Paris Arcades, describes their phantasmagoria, the affect-saturated projections of commodity spectacle.[19] In the nineteenth century, phantasmagoria were entertaining technical effects, ghostly simulations made by whirling magic lanterns that stimulated fright and awe. As phantasmagrams, the demographic transition model and calculations of averted birth were projections of a world composed of aggregate forms of life organized through rationalized interrelations, drawn into relief by indices that could serve as the levers for planning, each made by technoscience. They were felt projections built into the world that tended to erase their own conditions of creation. As a collective dreamscape, the conjugation of population and economy did not just come from a single model or equation but from a dispersed global cacophony of equations and simulations drawing in experts of many kinds and exceeding them. Technoscience dreams the world it makes sense in.

If we allow the premise that technoscience dreams macroeconomic atmospheres as it explicates them and that it conjures spectral counterfactuals as it counts, we might also ask, with whom does technoscience dream? With demographers, statisticians, and economists, yes, and so too nationalist planners, anticolonialists, and bureaucrats, and so too circulating postcolonial experts, Cold Warriors, and feminist reformers. The history of science has shown us that elite men have until recently predominated as technoscience's experts, and this was certainly the case with Cold War demography and its "Malthusian men," though women and feminists have in recent decades become able and ardent technoscientific agents of family planning and economic development.[20] Dreaming through so many discrepant political impulses, the phantasmic outcomes of GDP are perhaps so forceful because of the generative way they toggle between structuring the world as it is and imagining it otherwise.

In 1905 Begum Roquiah Shekhawat Hossein, a celebrated advocate for women's education and equality and an elite Muslim woman from what is now called Bangladesh, wrote a story about dreaming technoscience that is now considered the first feminist science fiction story.[1] At sixteen, Begum Roquiah married the deputy magistrate of Bhagalpur, who would die young. With the money left to her, she opened the first school for Muslim girls in India, which still exists today. In the portrait gallery of Dhaka's Pink Palace Museum, Begum Roquiah is the single female face looking out from the gallery walls.

Called *Sultana's Dream*, Begum Roquiah's story begins with the narrator falling asleep as she is "thinking lazily of the condition of Indian womanhood."[2] She then wakes into a dream. Alert within her imaginary, she finds herself walking unveiled in daylight in another world where the gendered structures of elite South Asian Muslim society are reversed: men are now confined to their chambers and women have become the scientists and public subjects. In this other world, "lady" scientists have turned away from building military machines and war. Instead they have invented ways to harness rain from the sky and share energy from the sun. One school of women scientists had "invented a wonderful balloon, to which they attached a number of pipes. By means of this captive balloon which they managed to keep afloat above the cloud-land, they could draw as much water from the atmosphere as they pleased." Another university had "invented an instrument by which they could collect as much sun-heat as they wanted. And they kept the heat stored up to be distributed among others as required."[3] The most distinguished science of this other world was botany. The roads were formed of a "soft carpet" of moss and flowers, and the city itself was a marvelous garden.[4] Sewing too was a celebrated art, and beauty highly valued. In the garden, "every creeper, every tomato plant was itself an ornament," such that the products of science were as much aesthetic as functional.[5] In this other world, science had a "sentimental" quality, not divorced from feeling or beauty.[6] In this world, women were all educated and married late, while men minded the children. Kinship was expanded such that "a distant cousin is as sacred as a brother."[7] It offered an alternative vision of how to live with plants and ecologies as part of the fabric of society. Revisioning gender, kinship, and science, *Sultana's Dream* conjures a feminist technoscience through a counterfactual world, a world turned upside down.

Awake to its own phantasy, the story calls on dispossessed women to dream with technoscience at a moment when South Asian Muslim women had begun agitating for access to formal education. Dreaming otherwise, the story draws on a tradition of Western utopian fiction and the revelatory force of dreams in Bengali culture to generate medical innovation and moral demands.[8] While the inversion of men's and women's worlds reverses rather than unravels gendered norms, the text suggests that dreaming is a domain where the possibilities of technoscience are contested. The inauguration of feminist science fiction as a South Asian Muslim feminist project disrupts the cartographies of knowledge-making that posits South Asian women as the objects to be counted rather than the subjects who dream with technoscience. Through its figures of gardens, clouds, and pleasure, *Sultana's Dream* suggests the possibility that technoscience might do life and phantasy differently.

The premise of *Sultana's Dream* has something to teach about the work of phantasmagrams. What would it mean to be awake, rather than asleep, to the dream? There is a politics in the invitation to become alert to the phantasy. How does one awaken to the phantasma of economy and population?

ARC II | Reproducing
Infrastructures

Who counts? Surveys and questionnaires pull crisp numbers and lines of correlation from everyday activity. In the new millennium, women and girls are figured in a swell of quantification and measures. Women's educational status and household wealth are positively associated with height.[1] Of ever-married women, 92 percent believe it is okay to refuse sex to their husbands if they have sexually transmitted diseases.[2] About 45 percent of households are exposed to secondhand smoke.[3] Half of girls aged fifteen to twenty-four do not have paid jobs, nor are they looking for work.[4] Girls are overwhelmingly aspirational about their employment prospects.[5] Possession of mobile phones has increased sharply, from 32 percent in 2007 to 78 percent in 2011.[6] Measures of body mass index (BMI), salt use and blood glucose levels, counts of TV and motorcycle ownership, statistics on pregnancies and sexual activity are all bundled by national sample surveys. Do you have a bank account? A loan? Clouds of data emanate from the busy work of social science. Ephemeral data about now, and just then, accumulate and the archive of data becomes crowded, filled with obscurantism and urgencies. This is postcolonial thick data.[7] Who and what is counted, who does the counting, and how do these numbers carry an affective charge?

All the figures and social facts listed above are reckonings about people from Bangladesh. While North American adolescents are intensively studied as the wellspring of lucrative markets, young women and girls in Bangladesh are among the most figured people in the world. They are favored research subjects of development. After a half century of exuberant social science projects studying women and girls, Bangladesh gives good data. On measures of the availability of data about girls, Bangladesh performs superlatively. For example, whereas the United States racks up a dismal 25 percent on a scorecard measuring available data about girls' access to income, Bangladesh scores an impressive 87.5 percent.[8] This, despite the fact that 78 percent of births of under-five girls are not officially registered; yet even this fact is another notch in the belt of counting.[9] National measures exist of the availability of toilets to households (34 percent), proportion who spend less than thirty minutes obtaining water (1 in 4), and percentage of women undernourished (24 percent) and obese (17 percent).[10] Overwhelmingly, numbers crowd to measure sexual activity, birth rates, family planning, and infant health from endless angles. Numbers are collected in national surveys, UN reports, or World Bank databases but also by local field stations and the thousands of NGOs that shadow the state. Numbers have been crucial to exposing murderous labor conditions, politicizing infant mortality, and recognizing pain and possibility. Numbers are needed at the same time that they are part of a larger non-innocent ecology of enumeration.[11]

Bangladesh glows with counting; it is audited and plotted not only exuberantly but excessively. Constellations of numbers extracted by national and global elites from the lives, livelihoods, and sex of others join GDP to thicken the technoscientific dream of governing population toward economy. While in Bangladesh electricity grids, water, or road infrastructures notoriously tend toward neglect, absence, and gridlock, the social science infrastructures of number making, in contrast, are robust, generative, and thick. Bangladesh is renowned among global social science institutions for the quality and quantity of its data.

All these numbers about the misery and aspiration of women and girls are numinous with feeling and imaginaries. They are linked to each other in a firmament of correlation to compose the greater phantasma of population. The numbers can help move donors, states, and NGOs toward particular investments and speculative futures. And more than that, the epistemic infrastructures of surveillance that generated such numbers were as much projects of statistical rationality as they were of affective rearrangement. Affect was at stake in the very act of making numbers, of counting, in conscripting research subjects to attach to promissories of "abundant life" when data was extracted from them.[12] Affect does not just animate numerical sums; it was purposively propagated by the infrastructures that do the counting. This is especially the case in the history of Cold War/postcolonial family planning that obsessively counted how people (and especially women) felt and behaved about fertility, sex, and birth control. This era of liberal transnational family planning built extensive social science infrastructures that aspired to stimulate the "acceptance" of contraception.

Bangladesh in the 1970s and 1980s was a particularly innovative site for the global development of new kinds of techniques for reordering population and economy together. While roads, pipelines, and power grids are considered the exemplary infrastructures of the built environment of modernity, I am suggesting here that the governance of population and economy in the second half of the twentieth century exuberantly proliferated enduring epistemic infrastructures elaborated out of social science research practices that also conditioned life. What infrastructures and conditions have been made out of the project of counting, family planning, and diminishing surplus life? Unlike the centuries-old tradition of the national census, postcolonial epistemic infrastructures associated with family planning were built to extract the numbers that at once gave form to the problem of population even as such research sought to intervene in the very phenomena under study. Disseminating birth control and intervening into flesh, such family planning projects sought to stimulate particular affective orientations and imaginaries thought to be conducive to reducing birth rates. Numbers do not only glow with feeling after the counting is done; social science practices surrounded the bodies that they counted with particular affective atmospheres.

Anthropologist Joseph Masco has brilliantly shown how stimulating collective affective states of fear and anxiety was crucial to the domestic U.S. Cold War military project. While the Cold War is so often chronicled as a period of game theory and worship of rationality, Masco's work shows how nuclear testing, films, war modeling, scenario planning, and civil defense exercises were projects intended to normalize dread, conjure threatening enemies, and distribute imaginaries of planetary destruction, an affective history that remains an enduring accomplishment after the Cold War.[13] A related history of conjuring affective atmospheres was at work in globalized U.S.-supported family planning, likewise underpinned by Cold War planetary commitments. Jackie Orr, in her study of panic in the nervous system of Cold War American infrastructures, names a politics of affect produced by, but also exceeding, the rational practices, spectacles, and structures of militarism, psychology, and social science.[14] Likewise, the affective circuits conjured by transnational family planning social science exceeded the calculative rationalities that organized them.

Frantz Fanon, writing toward the project of decolonization and out of his medical practice in the 1960s, provides further guidance to the charged multiplicity of affective politics within the conflicted dead ends of objectifying colonial science, imperatives to heal, the phenomenology of being dark skinned in a white supremacist world, and the affirmative aspirations of anticolonial projects.[15] Writing in the 1960s, Fanon traced how colonial medicine had profound affective consequence for Algerians in addition to the racist and prescribing mental health categories of the colonizer, such that sanity and pain became contested embodiments for knowing *and* not being known by colonization.[16] Fanon theorized how colonialism contradictorily moved through bodies and sensations. His work incites rationality, science, consciousness, feeling, and embodiment as all antagonistically caught in the promises and pessimisms of anticolonial politics.[17] Dense with phantasmagrams, the history of the economization of life, too, was formed through the contested evocation of affects that set up infernal contradictions.[18]

Summoning desires to alter oneself was the work of a transnational explosion of family planning surveys, instruments intended to both measure and provoke mass feeling.[19] In 1970 the Population Council published a large manual for the global standardization of Knowledge, Attitudes, and Practices (KAP) surveys.[20] A KAP survey captured pre-scripted answers characterizing awareness and "acceptance" of family planning. It wanted to count what women wanted. In 1970 more than four hundred KAP surveys already had been conducted in some forty-nine countries, renewing the global reach of social science after decolonization with a new set of techniques, problematics, and experts, now organized around intervening in life at the aggregate reproductive level. The KAP surveys were typically the first step in setting up a new family planning program; they were deployed as a quantitative demonstration of the existence of a local "desire" (and hence justification) for such an intervention, a numerical invitation to further meddling. In so doing, KAP surveys were postcolonial instruments that repackaged the longstanding politicization of the modernity of "women" and their "desire" as colonial, nationalist, and communitarian conflicts.[21]

Based on a sample survey method, KAP surveys are relatives of marketing studies that measure the likely interest of consumers toward a commodity.[22] As one demographer explained, "The most important function of such surveys is similar to any market research project: to demonstrate the existence of a demand for goods and services, in this case for birth control."[23] Julian Simon, a U.S. libertarian and economist, declared that "the KAP surveys are surely the largest worldwide market research job ever done," and that the global reach of such surveys was so profound that "in 25 years historians of science may assess this program as having a larger impact on humanity than any other scientific discovery or program, including the discovery of radioactivity or aspirin or bacteria or nuclear power."[24] Beyond measurement, KAP surveys as marketing devices were more importantly thought to *stimulate* interest in family planning, and therefore were deployed for their persuasive impact.[25] The KAP surveys helped to install infrastructures of persuasion as part of counting, and as the first act of family planning. They hailed desire while enacting its measurement.

The itinerary of such surveys traced the contours of American empire and its intersection with postcoloniality: first Indianapolis (1941), next Japan (1950), followed by Lucknow, Poona, and Mysore (1951), then Puerto Rico (1953), Khanna (1953), Jamaica (1956), Taiwan and Egypt (1957), and reaching Lahore by 1960.[26] Conducted between 1953 and 1960 in Punjab, the million-dollar-plus Rockefeller-funded Khanna study was among the most influential early field surveys in family planning. Designed at Harvard, it carved out its experimental field site from a "handful of villages" in the Ludhiana district.[27] Surveying required a physical assembly of World War II military jeeps, a fleet of bicycles, commandeered buildings, newly drawn maps, and standardized questionnaires, as well as a Punjabi staff of doctors, statisticians, coders, supervisors, and field-workers (both male and female), and a stockpile of condoms and foam tablets to be offered for free. The study visited homes door-to-door to interview residents, offered contraceptives, and tallied the results. It was lauded within family planning circles as a model study, copied globally for how to install the material organization needed for family planning research.

Under the auspices of the United Nations (UN) and with funding and direction from USAID, the project of installing family planning surveys became planetary with the ambitious World Fertility Survey (1974–87) that heralded itself as "the largest international social science research project ever undertaken," reaping a "rich harvest" of data tabulating the enthusiasm for contraception of some 350,000 people in forty-three countries.[28] Altering the fertility of poor people became a globalized U.S.-funded project in the name of preventing an apocalyptic future and of encouragement of capitalist sensibilities.[29] Family planning surveys exuberantly spread around the world. In South Asia there were twenty-eight KAP surveys already by 1970.[30] One researcher claimed that there were as many as four hundred family planning studies conducted or underway in India alone by 1973.[31]

Whatever the exact count, South Asia after partition had by far the densest regional concentration of surveys, signaling the effusive social science infrastructure built in India, Pakistan, and later Bangladesh out of the problem of sex.[32] Decades of local birth control activism, state agencies of quantification, and ensconced routines of bureaucracy converged into the postcolonial study of "population."[33] For Americans, with the help of popular books and media, South Asia was imagined through the rubrics of crowds, mouths, and bombs.[34] So too was South Asia singled out by U.S. foreign policy as a site of threatening overpopulation. In the now declassified 1974 report *Implications of Worldwide Population Growth for U.S. Security and Overseas Interests* commissioned by Kissinger, Bangladesh, India, and Pakistan were listed at the top of a priority list of "countries where the imbalance between growing numbers and development potential most seriously risks instability, unrest, and international tensions."[35] While the document extolled "the universal objective of increasing the world's standard of living," at the same time the report prophesied a dystopic future that would unfurl if the United States did not intervene into global population, a world characterized by "high and increasing levels of child abandonment, juvenile delinquency, chronic and growing underemployment and unemployment, petty thievery, organized brigandry, food riots, separatist movements, communal massacres, revolutionary actions and counter-revolutionary coupe."[36] The assessment concluded that "the Subcontinent will be for years the major focus of world concern over population growth."[37]

In the 1970s, then, there was an intensive encounter between U.S.-supported and regional family planning programs that propagated survey practices as a kind of governmentality of population, a technique for simultaneously measuring and rearranging the affective orientations and living-being of aggregates of people. The KAP surveys would help to stimulate "population," and not just economy, as a felt condition. By 1970, a KAP survey would use the "full market approach," which highlighted socioeconomic measures of consumption, motivation, wage-labor participation,

C. Ownership of modern objects

> NOTE: The following list starts with objects that will be relevant for underdeveloped societies and goes on to include objects that are found only in somewhat more developed countries. This list should be adapted to the particular country being studied.

1. Do you have electricity available in your home?

       Yes_____   No_____

2. Do you own:

	Yes	No
A clock or watch	—	—
A bicycle	—	—
A radio	—	—
A sewing machine	—	—
An electric fan	—	—
A motorcycle	—	—
A motor scooter	—	—
A refrigerator	—	—
A car	—	—

5.1 An example of a KAP survey question used in Taiwan with pre-scripted answers about economic attitude, thereby measuring and stimulating a "conscious calculus of choice." (Population Council, *A Manual for Surveys of Fertility and Family Planning*, 1970) propensity to save, and so on.[38] More than this, the "full market approach" was also about drawing out and measuring desire, the "impressionability" of that desire to mass media, and the relation between feeling and its commodity fulfillment. Should fertility be up to fate? Is prosperity due to luck, or hard work? In short, KAP surveys explicitly sought to provoke, alter, and quantify aggregate "attitudes" as they were aligned or nonaligned with marketable desires and economic futurity.

Once surveyed, a sample might well be surveyed again and again, marking change over time in response to different family planning and development programs, thereby fostering the sample as a longitudinal experimental site.[39] Surveys were not snapshots; they were provocations toward modifying and scripting the behaviors or feelings they tracked. Surveys desired desire; they called on subjects to want contraception, and experimentally attempted to trigger that want. Refracting fertility through attitudes and consumption, KAP surveys were but one of a profuse array of social science techniques spreading in the 1960s and 1970s that sought to transform fertility by stimulating contraceptive "acceptance."

"Attitudes" had been a concern of some of the earliest and formative family planning survey work. Frank Notestein's first research project, for example, was a study of women at the Margaret Sanger Clinic in New York during the 1930s, from which he concluded that it was not the availability of birth control that created fertility declines but the acquisition of modern attitudes.[40] This argument was elaborated by demographer Ansley J. Coale and his economist co-author Edgar M. Hoover in their landmark *Population Growth and Economic Development in Low-Income Countries* (1958), which influentially laid out the case for why reducing population was necessary, and not just incidental, to future economic development. Coale and Hoover named the acquisition of a "calculus of conscious choice" as a prerequisite for fertility decline, thereby encouraging the inclusion of questions that sought to simultaneously hail and detect numeracy, consumption, expectations about the future, and facility in banking practices as attributes of the modern subjectivity necessary for family planning.[41] The "calculus of conscious choice" was as much defined by proper feeling as it was by correct rationality.

By the 1970s, USAID was the largest funder of family planning projects globally. Measuring and then fulfilling "unmet need," defined as the gap be-

Bangladesh – 24% of women do not want pregnancy but are not using contraception

The unmet need for
family planning

5.2 A typical portrayal of "unmet need" featuring a young woman from South Asia as the iconic subject of this affective condition. (UNICEF, "The Unmet Need for Family Planning," 1995)

tween the desire for fewer children and the availability of contraception, was at the crux of USAID's abundantly funded family planning policies. The agency developed a photographic vocabulary of feeling toggling between unmet need and happy consumption, whereby brown, poor, often Muslim women would choose between suffering want (figure 5.2) and smiling fulfillment with contraception. Contraception, thus, was offered as a cluster of (impossible) promises, not only of rationally controlling one's body but also of the good life, of satisfaction through capitalism, of joining with national aspiration and enhancing collective wealth.[42]

For Reimert Ravenholt, the director of USAID's Office of Population

(1965–79), the feeling of "unmet need" was something that could be stirred by the very presence of the object of its fulfillment. According to USAID's preferred "supply-side" method to the distribution of contraception, rapid contraceptive uptake in a country did not require elaborate development schemes that would change the fundamental conditions in which people lived.[43] Rather, the very act of making available family planning technologies magnetically worked to compel people to accept them, whatever the circumstance. Ravenholt was inspired to develop his "supply-side" approach by drawing an analogy to the distribution of Coca-Cola:

> If one established a few places in a country where Coca Cola could be purchased, at a considerable distance from where most people live, at uncertain hours, necessitating that buyers have a sophisticated knowledge of the distribution system, no doubt the sale and use of Coca Cola would correlate considerably with general education, occupation, economic and transportation circumstances, sophistication, etc. But if one distributed an ample free supply of Coca Cola into every household, would not poor and illiterate peasants drink just as much Coca Cola as the rich and literate urban residents?[44]

Notably, USAID's appeal to an iconic commodity of globalized American capitalism and its logistical distribution did not rest on the straightforward assumption that convenience increased consumption. More subtly, the supply-side approach held that the very act of offering contraceptives aroused demand where it might not have been before. Ravenholt demonstrated his theory at a keynote lecture he gave at a population conference in 1976. He first polled the audience to see how many people wanted a cookie. Few expressed an interest. Then, he staged an intermission during which half the audience was actively offered cookies as they left the room; 96 percent accepted a cookie.[45] If KAP surveys measured unmet need as the gap between desire and the availability of supply, Ravenholt's supply-side technique reoriented unmet need as a provocable desire, and hence demand, that could be prompted by supply. Presenting someone with a commodity/contraceptive was a means to incite the desire that surveys wanted to count.

The KAP surveys were merely one of a suite of family planning techniques intended as universally applicable methods for hailing, altering, and satisfying mass *affect*—capacities to feel, think, and desire, or, put another way, capacities to respond. The surveys encouraged a particular "affective economy" of heterosexual nuclear family units whose intimacy and consumption was to be governed as desiring subjects, as subjects oriented within the promises of abundant life.[46] Geeta Patel suggests that the advertised and ubiquitous radiant happiness of family planning propaganda in India performs as a kind of "hundi," a circulating exchange in which the payment due for acquiring heterosexual propriety keeps moving between hands and sites with the promise that it will someday arrive. It is a promise that can be circulated and exchanged, though its arrival can be perpetually deferred.[47] Happiness and future national economic prosperity was an unpayable debt in exchange for consuming family planning technologies and attaching to circuits of heterosexual propriety and consumption.

In family planning, consumption and experimentation would meet at a threshold of indistinction. Family planning field-workers simultaneously gathered data, sold promises, and offered contraceptive services for consumption within the very same encounter. Studies tweaked scripts and practices looking for more efficient choreographies of the moment of acceptance of a condom, a foam tablet, an IUD, a package of pills, an injection, or even sterilization. In this liberal setup, to be experimentalized was to be induced to voluntarily "choose" to participate from a limited set of possibilities, to decide to consume what was on offer. Surplus population was to be combatted with consumption. Consumption was the desired experimental result. Consumption merged with becoming experimentalized.

The connections between KAP surveys and marketing techniques, between experiment and consumption, were structural and explicitly crafted. The KAP surveys would become pivotal to the widely used practice of "social marketing," in which advertising is used to instigate a behavioral change and the market is used to disseminate subsidized commodities to stimulated users. Social marketing as a neoliberal technique of development was first implemented in 1964 by experts at the Indian Institute of Management in Calcutta who designed a campaign to distribute the Nirodh condom.[48] The KAP surveys guided the oldest and largest transnational social marketing organization, Population Services International (PSI), an NGO started in 1974 by two Americans with a North Carolina sex toy business and funded by USAID. The first PSI campaign was in Bangladesh, where it distributed the "good life" of Raja condoms and Maya birth control pills, which were later joined by life-saving oral rehydration therapy (ORT), as USAID-subsidized commodities sold by small shopkeepers across Bangladesh.[49] In social marketing, the work of measuring and soliciting desire was amplified by the persuasion of advertising and the market as a purportedly efficient commodity distribution system. The contraceptives themselves were sold at below market prices, often close to free, but nonetheless still circulated as commodities.

At the heart of this approach was the belief that open markets not only offered the best distribution system for family planning supplies (and later public health interventions) but also for stimulating acceptance and affective attachments. The metric of a project's success or failure was the number of units "accepted," imagined as both an emotional and physical act, with the expectation of a corresponding decrease in birth rates later. "Acceptance" was expected to translate to less aggregate lives and a corresponding reduction in the denominator that calculates GDP per capita. "Acceptors" were thus counted with the expectation that averted births would follow.

My point here is that the ambit of techniques that make up the economization of life includes not just the register of macroeconomic entanglements with reproduction via state population policy and modeling but also an abundance of surveys and marketing as methods of stimulating consumption within experimental modalities. Fertility, affect, and consumption became an emergent "field site" toward the mass prevention of births, and indirectly for governing the economy. The KAP surveys, supply-side family planning, and social marketing were symptoms of a metamorphosis in the history of postcolonial governmentality associated with the economization of life in which the grandiose state planning of the 1950s and 1960s, with an emphasis on demographic transitions, population policies, and targets at the national level, was transformed into the proliferation of more neoliberal, nongovernmental experimental projects that brought client populations into circuits of counting that not only provided access to family planning commodities and services but also invented new ways of harnessing feeling and living-being to projects of fostering the "abundant life" of modernization, which was also the surplus of life better-not-born.[50]

The KAP surveys materially manifest the historical importance of affect to the economization of life beyond the palpable phantasmic horizons of macroeconomy and population. They might well be understood as evoking mass affect as a target in need of adjustment through commodities, not for the sake of profit directly linked to their consumption but for the sake of the end goal of the macroeconomy and installation of a particular affective milieu in which capitalism operates.[51] This is not to argue that attitudes and desires were implanted in subjects but rather that particular infrastructures of mass affect were fostered in "economies" quite literally.

Despite the planetary propagation of family planning studies, from their very beginning many were seen as failures. The failure to alter behavior even when "acceptance" was recorded was a finding that experts bemoaned at length. Almost as soon as KAP surveys were deployed, a "KAP gap" was declared. A KAP gap was the disconnect between general favorable attitudes (acceptance) toward family planning and actually using contraception. The KAP gap troubled what it meant for someone to be an "acceptor." Someone might take a package of pills at the door, but would they actually consume them? They might view family planning favorably in a survey question, but would they actually rearrange their sexual practices, their aspirations, their bodies? How does one close this space between feeling and doing? How is it possible to build infrastructures aimed at inciting and motivating people to action for the sake of less future life?

The famous Khanna study that helped to establish the protocols of field site survey taking was one of these failures, revealing the gap between "acceptance" and reducing birth as a kind of negative finding. Lasting eight years, and followed up again a decade later, the Khanna study built an enduring research infrastructure to enable house-to-house surveillance intended to stimulate people to accept contraception. Obsessed with acceptance, the study was contradictorily indifferent to the actual affective charge of the survey for the research subjects.[52] No corresponding change in birth rate accompanied the counts of acceptances logged.

A young Harvard anthropologist, Mahmood Mamdani (who would later become an eminent expert on East Africa and violence), revisited the site of the Khanna study in the summer of 1970. He would draw on this research to write *The Myth of Population Control*, accusing the Khanna project of being an imperialist sham.[53] Interviewing the study's past research subjects, he argued that what counted as "acceptance" of contraception was in fact a form of hospitality. People, in gestures of kindness and politeness, endeavored to tell their interviewers what they wanted to hear: "Babuji, someday you'll understand. It is sometimes better to lie. It stops you from hurting people, does you no harm, and might even help them."[54] They further explained, "But they were so nice, you know. And they came from distant lands to be with us. Couldn't we even do this much for them? Just take a few tablets?"[55] Boxes of foaming tablets were stored under cupboards, tucked under beds, or even repurposed as materials for art.

The "acceptance" that researchers believed they were provoking and measuring was, to a large extent, their own phantasma. The attachments to population, abundant life, and the calculus of conscious choice circulated through the field-workers, the professional staff, the elite donors, researchers, state officials, cooperating businessmen, and the experts, rather than the research subjects. One of the most potent results of the experiment, thus, was about the experts themselves, the building of a research infrastructure, and the conjuring of an extra-state experimental modality for family planning that was attached to consumption. What the surveys accomplished was a particular kind of assembly—populated with forks of choice, modernity, persuasion, desire, consumption, and promises—as the surround for the neoliberal problem of population tethered to economy.

The history of how social science practices summoned affective atmospheres has something to teach about subsumption, that is, about surrounding life with the forms and phantasies of economization. *Subsumption* was a term Marx used to name the way arrangements of work, forms of life, and social relations were absorbed into capitalist relations.[56] To subsume means to bring under something larger, which in Marxian thought is typically capitalism writ large. Torquing this formulation, I suggest that at stake in subsumption is not merely the absorption of forms of life into a larger set of already given relations but also the historical formation of the very galaxy of those relations, the making of the surround itself, the creation of the atmospheres and assemblages that capitalism conjures for itself and as its own context. If the macroeconomy became the spectral firmament of the nation-state, and population the aggregate figure that dreams the totality of life in a nation, we can think of projects to rearrange the "acceptance" of populations as projects to summon and assemble the atmosphere that particular projects of capitalism demanded as their surround.

If surveys and marketing are not enough to avert birth, how does one build the environment of motivation required to incite people to choose to re-duce their fertility? Family planning experts agonized over this question: Was facilitating people to have the number of children they wanted enough to meet targets and thwart the population bomb? What other measures might be necessary? Could a context of motivation be built that would push people to take up birth control at higher rates?

Motivation—as a communicable incitement to action—would become another buzzword of family planning.[1] How is it possible to create condi-tions of continuous motivation toward reducing birth? Field-workers were now called "motivators." Programs were to build up their "motivation-delivery capability."[2] Effective motivation required "change agents" with "homophily," field-workers similar to the people to be motivated with the exception that they already used and promoted birth control.[3] Billboards, brands, and advertisements alone would not do. Motivation in the realm of the personal propagated best in person, with your neighbor, door-to-door, face-to-face. Field-workers must be local and gender matched. Vasec-tomy canvassers should have had vasectomies; lady field-workers should use the IUD. Their own bodies were to be part of the infrastructure of per-suasion. Motivation was to pass between bodies, become contagious and collective.

Beyond door-to-door dissemination, opportunities for motivation were phantasized as everywhere: each encounter with a state agent or a shop-keeper was a chance to conduct motivation. The entire Bangladesh family planning program was described by one U.S. expert as an apparatus of moti-vation: "a good deal of the energy of the apparatus that is being built up is destined to provide not services proper but 'motivation.' In other words, the system is to *create* the demand that subsequently it is supposed to satisfy."[4]

Importantly, motivation was to be amplified with incentives. Motivating agents could be incentivized with payments, as could accepters. The Ernakalum sterilization campaign in Kerala, India, was held up as an example of what continuous motivation might accomplish. Everett Rogers, the influential communications expert whose work helped to elevate motivation as a concern, praised it as an "all out effort" in which every public servant was ordered to drop their regular duties to promote the campaign. Medical personnel, tax officials, publicity officials, and train ticket sellers were all to motivate vasectomy adopters to be sterilized at a mass vasectomy camp, itself promoted with a "festival" ambiance.[5] Each successful promotion earned a motivator ten rupees. Each female acceptor was incentivized with more than forty rupees in cash, three kilograms of rice, a sari, and a lottery ticket.[6] Money and food became motivating agents. Described as a "total government effort, involving every ministry, not just those officials regularly assigned to family planning," the Kerala sterilization campaign gave form to the phantasy of a total motivation (akin to the total societal mobilizations of World War II) toward altering population.[7] This experiment was the precursor to the coercive mass-production sterilization camps of the state of emergency in India, inaugurating an era of population control that filmmaker Deepa Dhanraj called "Something like a War."[8]

The phantasy of total war also shaped Pakistan's first two decades of population control planning starting in 1965. Beginning with Ayub Khan's enthusiasm for population control, Pakistan's program dreamed up a project of household-level motivation, amplified by the offering of door-to-door services, through a new corps of 1,392 family planning officers, 1,200 lady family planning visitors, 1,209 family planning assistants, and 25,000 part-time dais (traditional birth attendants) who were then paid referral fees for each IUD acceptor they secured.[9] With Ayub Khan's downfall in 1970 (and with the violent split between Pakistan and Bangladesh soon after), a new family planning administration declared a program of "continuous motivation" that sought to triple expenditures on family planning and replace dais (now criticized as illiterate and ineffective motivators) with male and female motivation teams. A network of some 50,000 shopkeepers were planned as a conduit toward contraceptive "inundation."[10] In practice, field staff were hard to hire, or became positions of political appointment used toward political canvassing. Shopkeepers were reticent to sell. Records were not kept or were fraudulent.[11] Total motivation was a mirage.

In attempts to choreographic continuous motivation, U.S.-funded studies habitually identified financial incentives as an oil that could lubricate motivation toward family planning.[12] Economist Stephen Enke had long been a promoter of financial incentives as a potent tactic that he simulated in his population models.[13] USAID held that incentives were "economic rewards to induce people to 'recalculate' their desires for additional children."[14] Direct monetary payment was coded not as a form of exchange or pay for labor but as a means of incentivizing, of inducing enthusiasm and acceptance. Financial incentives could motivate the supervisors, the nonprofessional canvassers, and the accepters alike. Money as an inducement became part of the assembly of family planning. Put at a high enough rate, "acceptance" numbers could dramatically shift. Thus, the efficient rate of incentives became its own calculative problem. The precarious, the dispossessed, and the old, it turned out, could be persuaded by the thousand and even the million for mass sterilization conducted at "camps" when the payoff was large enough and when sterilization was the only option paying.[15] The choreography of "choice" was intensified as a mass campaign of financial manipulation.

For the liberal political imaginary being summoned, subjects were free to choose to trade their fertility for pay. Incentives do not force but merely stimulate. In rejoinder, one might instead note that stimulation itself was a kind of coercion. Monetary stimulation joined other kinds of coercions that were there to some degree all along, in the choreography of prepackaged answers, in the small range of contraceptive options, in taking advantage of hospitality, in the projects of counting that were intended as persuasions. The environment for deciding on fertility was summoned as a matrix of incentives and disincentives, an imaginary of push and pull forces that planners could intervene in. In the prestigious journal *Science*, Lenni Kangas, the deputy directory of USAID's Office of Population in India, proposed a total motivation scheme that he described in the hypothetical "Village of Malthusia." In this scheme, community development and incentives would be merged: communities would be rewarded with tube wells, power generators, or resources to build an agricultural cooperative if enough members remained free of pregnancy that year. Withholding and rewarding economic development itself—termed "integrated incentives"—was being suggested as the kind of "bolder experiment" that reducing population growth required.[16]

Bernard Berelson, an American behavioral science expert on voting and president of the Population Council (1968–76), influentially posed the problem of imagining family planning *beyond* the technical act of contraception, searching for the line of incentive use (and hence liberal politics) just short of state enforcement: "Where does effective freedom lie? With the free provision of information and services for voluntary fertility limitation? With that plus a heavy propaganda campaign to limit births in the national interest? With that plus an incentive system of small payments? large payments? finders' fees? With that plus a program of social benefits and penalties geared to the desired result? Presumably it lies somewhere short of compulsory birth limitation enforced by the state."[17] Just short of the compulsory, new assemblies of incitement could be planned.

Bangladesh, too, had a financial incentives program for sterilization that began in 1976 and intensified in 1983 under the military dictatorship of Hussain Muhammad Ershad. A new two-year "emergency program" was mounted under the pressure of donor agencies, including the World Bank, the UN Fund for Population Activities, and USAID (which funded the vast majority of the program).[18] The incentive program was accompanied by a larger project of building an infrastructure of extensive motivation: every ministry was charged with the task of arousing sterilization acceptance, including the army.[19] Two feminist activists and scholars, Farida Akhter in Bangladesh and Betsy Hartmann in the United States, have documented the results in detail. Much like the Ernakalum campaign, people who were sterilized would receive money (as much as a few weeks' wages for some), a sari or lungi, and a card that gave access to food relief. Perversely, without the card, destitute people were denied food.[20] The sterilization rate spiked before the harvest, when hunger intensified.

In the spread of incentive programs, the plan of mobilizing marketing as a conduit for the effective diffusion of family planning was mutated into a project of trading monetarily on the survival of precarious people, of directly paying people to alter their bodies, their sex, their futures in the name of a greater rate of return for the nation that would come from the quantitative drop in their future number. This payment, however, is not named as an exchange, or as a wage, but merely as a stimulant. The objectifying factory conditions of the mass sterilization camp, the disregard of follow-up care, the paucity of informed consent, the lack of mortality statistics collected following sterilization, the absence of facilities to sterilize surgical instruments or to even glove the physician's hand—all pointed to the general contempt for the lives of those targeted for sterilization, to their disposability in the present, and not just in the future.[21]

Incentive schemes had become integral to the liberal experimental assemblies that constituted transnational family planning. Each project of stimulation was also an experiment, propagating more numbers for the next round of motivation. Cold War/postcolonial social science had miraculated population in a web of relations, potentials, and incitements that were now open for experiment. While Raymond Pearl had inaugurated the experimentalization of population in his jar of fruit flies, globalized family planning had spread experiments into the planetary scale.

Experiments thrived in the new state of Bangladesh founded in 1971 in the aftermath of disaster, war, and famine. Trials, training, and pilot projects multiplied across rural space: agricultural schemes, loan programs, village self-sufficiency, immunization campaigns, as well as an abundance of family planning ventures. In 1970s Bangladesh, calls for the reordering of life abounded: saving, lending, averting, uplifting, preventing, educating, empowering, training, disseminating, incentivizing, immunizing, motivating, inundating. A thick history of social science in South Asia, of colonial enumeration, of bureaucratic culture, of village governance, of socialist planning, of community organizing, as well as of U.S. imperialism was reactivated into a new era of experimentality.[1]

In 1978 Bangladesh's Ministry of Health and Population Control convened a national conference in Dhaka to review the profusion of "innovative projects in family planning." Some were state projects, yet many were conducted by "voluntary organizations," variously local, national, and transnational.[2] The altering of population was an urgent national priority that ambitiously aspired to "maximize local participation" and reach every woman in every village.[3] "Voluntary organizations need flexibility in their operations. No effort will be made to unduly control them," the Ministry declared.[4] Today, the archive is thick with hundreds of family planning studies conducted in Bangladesh in the 1970s. Each project had staked out its own experimental terrain: 101 villages here, 600,000 over 6 thanas [districts] there.[5]

Some were self-reliance projects or Mothers' Clubs oriented toward a self-help horizon, with family planning merely one axis of intervention. Others began as relief projects responding to earlier flood and famine, now appending family planning programs in the aftermath of crisis, stimulated by the torrent of funds from the United States urging population control. Most were door-to-door projects featuring the mobilization of women field-workers as "motivators" distributing contraceptives or urging sterilization. Some projects insisted on prioritizing immunization and infant health, diverting funds from averting births. Yet others were narrowly focused on the logistics of increasing acceptors through mobile mass sterilization camps, or on social marketing campaigns of "inundation" to put condoms and pills in easy reach on shop shelves. Experimental assemblies seemed to breed more experiments, each a little different. In the name of reducing birth, experiments themselves were what reproduced.

Over the 1970s, experiment, as a kind of governmentality, became a ubiquitously yet unevenly distributed form in Bangladesh. Experiments offered an infrastructure of off-the-grid governmentality, a mode of assembling practices toward survival and a reordering of life in Bangladesh's many sites of past violence, particularly where the state was dysfunctional or where large social-technical infrastructures of industrialization and modernization did not reach.[6] Colonial rule had long approached colonies as living laboratories for extractive scientific practices.[7] But something different was happening in the 1970s. Within the new practices of development, tangled in postcolonial national modernization projects as much as emergent forms of transnational imperialism, experiments transitioned from a practice within other institutions to the very assemblage of governmentality itself. Experiments had stretched out beyond the lab or the clinic, or even the field site. Bangladeshi experts and local leaders themselves propagated experiments at the juncture of local politics and transnational inducement. In the second half of the twentieth century a *spatialization of experiment* occurred, inhabiting markets, connecting villages, giving rise to new kinds of expertise and labor.

Anthropologist and HIV physician Vinh-Kim Nguyen has called this a kind of *experimentality*, in which the provision of services, health care, and support does not come from a centralized state infrastructure or from the market but from a distributed field of experiment coordinated in local-transnational-NGO assemblies.[8] As Nguyen has shown in his work on the "massive multi-sited laboratories" of HIV care in turn-of-the-twenty-first-century West Africa, entering into transnational experimentality has sometimes been the price of survival. Moreover, experimentality, as a mass intervention into consciousness and bodily practices, relentlessly produces evidence that then legitimates continued interventions as a self-perpetuating relation of rescue. There is a vibrant scholarship in anthropology on experimentality that I build on for this history of experimental exuberance in Bangladesh.[9]

Experimentality as it emerged in the 1970s contributed to building the infrastructures and epistemologies that made up the economization of life. And further, the history of family planning in the 1970s is crucial to the emergence of global health experimentality later. For example, HIV care in the 1990s was often initially assembled through already existing infrastructures of "reproductive health" care, in turn built out of the family planning infrastructures of the 1970s.[10] The habitual study of "experimental populations" in 1970s family planning would expand to become a more generalized mode of experimental governmentality. In Bangladesh and around the world, experimentality became a mode for connecting life chances to economy through interventions into sexed bodies. The endless futurity of experimentality became an extensive infrastructure for the mass conjuring of the dreamscapes of economy and population.

What is an experiment? One might be tempted to define a proper experiment narrowly on the model of a clinical trial with its use of randomized control and test populations. Experiment, however, is a more generalized and loose mode inclusive of many forms and ideologies. Experiments are found in both the arts and sciences, as well as practices of the self. Science studies scholar Hans-Jörg Rheinberger describes experiment as an ongoing "process of reorientation and reshuffling of the boundary between what is thought to be known and what is beyond imagination."[11] In family planning, experiments tend to be oriented toward instrumentality and intervention, not surprise. Hence, I define *experiments* in a broad sense as technical-social assemblies that arrange and gather data about interventions into the world toward the possibility of making something different happen. In this sense experiments are conjectural future-making assemblages.[12] In experimentality, access to different futures can be conditional on attaching to experiments.

What was most generative about the experimentality emerging through family planning in the 1970s was not the production of certain kinds of pragmatic results (for many experiments had meager results) but rather the building of infrastructures positing aggregate life as a recomposable conjectural domain open to and in need of repeated intervention. Family planning experimentality sought to recompose the very relations that made up population, that made up families, choice, sex, gender, and poverty—the relations that assembled the atmosphere of aggregate life. Experimentality hails life as composed of potential, of chances, of possibilities for becoming, of manipulable relations that can be triggered and altered, if only the right protocol and technique can be deployed. The aims of reordering relations took many political forms: coercive and empowering, careful and negligent, feminist and entrepreneurial, improving health care or narrowly focused on preventing birth. These differences between kinds of experiments mattered profoundly to the lives and life chances of the people implicated in them. At the same time, the promiscuous spread of experiments in so many various alignments signals a broadly distributed condition of rendering the world as open to remaking in a general sense.

Experimentality shared with capitalism this investment in conjuring worlds of potential open to recomposition and speculation. Experimentality constituted a milieu amenable to capitalism's desire for changing and changeable surroundings. Experimentality in Bangladesh did not necessarily directly produce new forms of property or privatization but rather stimulated assemblies of individualization, choice, motivation, gender, desire, quantification, confession, and risk as a recomposable milieu, as an environment, as a surround. Even if nothing was improved, the larger surround was still miraculated as a firmament with yet untapped potentials. In this way, experimentality could function as a form of subsumption, that is, of surrounding life with the forms and phantasies of economization, as well as the instruments and infrastructures of expectation.

Crucially, experiment is not aligned in this way a priori; rather, there are many ways of doing experiment. Invested in provoking an otherwise, experiment can be part of decolonial and other radical projects. This is critical. I am committed to the potential of experiment for generating life otherwise, for yearning toward the possibility of other worlds and other arrangements that might be less violent and more affirming to life. Pragmatically, too, participating in experiments can be an individual act of care, coping, collectivity, or aspiration, as many ethnographies of family planning and global health have shown.[13] Participating in experiments can be part of making do, as well as self-preservation when options are highly constrained. When living in hostile conditions, self-care experiments can be an act of "self-preservation, and that is an act of political warfare," as Audre Lorde teaches.[14] Moreover, the extensive experimentality of family planning is not singularly oriented toward subsumption. Feminist politics, for instance, has consistently taken part in this experimental exuberance. For example, the Bangladesh Women's Health Coalition has innovated ways to offer reproductive health. Self-help and cooperative projects were an important strand of experiment within 1970s Bangladesh.[15] So too is there a history of feminist research concerned with collaborating with women about their experiences as research subjects that has challenged dominant ways of doing experimentality.[16] All sorts of people and collectivities are able to strategically access life-affirming assemblies through experiment, from women seeking safe abortions, to the availability of life-preserving medicines, to material supports for reducing the risk of maternal mortality, to the obtainability of clean water. Experiments are animated with mixed and contradictory intentions. Yet even though experimentalities can and do bring affirming material possibilities into people's lives, I am suggesting that there was an overriding hegemonic form of experimentality that was invented in and recomposed 1970s Bangladesh.

Regardless of good intentions or individualized benefits, experimentality is always accompanied by destructions and losses. Experimental exuberance calls social relations into febrile rearrangement, legitimating a continuous refreshing of destruction at a microscale as relations are offered up as decomposable and called to recomposition over and over. Once something is decomposed, it can rarely be returned; there is no going back. It can only be recomposed again, and again. Every effort to do life differently, to invent the better intervention, also serves to intensify this sense of openness to mutation. Talal Asad captures this process, stretching across colonial and postcolonial governmentalities, in which "new possibilities are constructed and old ones destroyed," such that the changes "do not reflect a simple expansion of the range of individual choice, but the creation of conditions in which only new (i.e. modern) choices can be made."[17]

Thus it is crucial to understand that the experimental exuberance in Bangladesh came after a century of destructive alteration: a colonial British occupation, a 1905 partition of Bengal into East and West only to be rejoined again, a hard-won decolonization in 1947 tied to another violent partition further complicated by the drawing of a new nation-state of Pakistan with two geographically unattached wings of East and West Pakistan (with the capital in the west, creating another configuration of rule at a distance), and then independence wrought in violence in 1971.

The scope of mass death in Bangladesh between 1970 and 1975 is difficult if not impossible to represent. Numbers are pivotal to how this period is memorialized and contested. The counts themselves are unattainable projects, impossible reconstructions, that nonetheless are necessary to fathom the scope of loss or make human rights claims.[18] Before the war even began, a monstrous storm, the Bhola Cyclone of 1970, had, in the storm surge of a single morning, ended some 300,000 lives (the true death toll will never be known).[19] It was the deadliest hurricane ever recorded. Bengali-speaking East Pakistan was geographically remote from Urdu-speaking West Pakistan, which held the seat of government, then militarily controlled and headed by General Agha Mohammad Yahya Khan. A sense of West Pakistan's indifference to the disaster fueled East Pakistan's independence movement. A month after the cyclone, the East Pakistan–based Awami League won control of the parliament as well as the prime minister position, but West Pakistan leaders refused to concede. An independence war followed, pitting Bangladesh militia against the might of Pakistan's Cold War–funded military. Only with India's intervention was the war won in Bangladesh's favor. Deaths from the war touched everyone. Intellectuals, doctors, political leaders, engineers, and other professionals were killed by the Pakistan army; professors and students at the University of Dhaka were notably targeted, as were Hindus and non-Muslims. Bengalis in turn persecuted Urdu-speaking Biharis. The war took a genocidal form.[20] War rape by the Pakistan army was systematic and widespread, enflamed by the racialization of Bengalis as Hinduized impure Muslims.[21] Houses were burned, and the dead piled in mass graves or thrown into rivers. How to count the dead remains a point of dispute: at the time, Pakistan argued only twenty-six thousand were killed, while the new state of Bangladesh claimed three million dead.[22] Scholars today try to reconstruct and estimate the count with censuses, newspaper reports, and mass-grave forensics.[23] But the very act of critically investigating the production of official numbers can lead to contempt charges.[24] Adding to this disaster, millions were put into motion. Cholera, which loves nothing more than a refugee camp, snatched more lives. The round number of ten million is typically given for the number of refugees fleeing to India, a staggering tally, in which individual lives are obscured from view and only the sheer massiveness of loss is conveyed. One Indian state document puts the count at the exact figure of 9,889,305.[25] This refugee crisis was the largest single mass displacement of people in the second half of the twentieth century.[26] By 1975, famine and cholera had killed as many as one and a half million people *in addition to* those dead from the war and typhoon.[27]

The force of the famine after the war was fueled by the decision of the United States under President Richard Nixon to withhold food aid in part tied to demands for Bangladesh to implement U.S.-requested Cold War policies about Cuba.[28] Amartya Sen's work arguing that the famine was caused not by a lack of food but by hoarding, speculation, and inequalities in food distribution would later help to win him the Nobel Prize in Economics.[29] Mass death had become a problem space of calculation, of economic models, of excess mortality estimates, and of politics. While the exact number of war deaths is heatedly debated, and arguments continue over whether this violence qualifies technically as "genocide," this period of Bangladesh's history stands out along with the European Holocaust, Rwandan genocide, systematic killing under Stalin, and viciousness of the Khmer Rouge as among the deadliest achievements of the twentieth century. From 1970 to 1975, then, an intensive conflagration of late twentieth-century forms of mass destruction had reshaped lives in Bangladesh: a climate-based natural disaster, a genocidal war, war rape, a refugee crisis, and an administrative and market-produced famine.

Bangladesh was thus forged in this violent crux. The extremity of destruction and the feebleness of the emerging state created a condition in which experiments acquired a moral force. How was it possible to go forward?

Mass death, not dying, not being born: these would become three necropolitical modes composing Bangladesh in the late twentieth century, and it is necessary to trace them together to understand the economization of life and experimental exuberance. In using the term *necropolitics*, I draw on the work of Achille Mbembe, who argues that there has been a modern instrumentalization of the material destruction of human bodies and populations in the form of concentration camps, mass internment, occupations, refugee camps, apartheids, war, suicide bombing, and other coordinated exposures to violence and terror.[30] In necropolitics, death and exposure to death are turned into productive ends. In the history of Bangladesh, the mass destruction of 1970–75 was productive of experiments as a way of calling forth better futures.

With bold visions of a "socialist transformation of society," and yet humbled by the extent of damage and "decades of neglect," the Planning Commission's first Five Year Plan (1973–78) describes "development" as "a long and painful process. It means present sacrifice for future gains." Despite all the war loss, population growth is still identified as a key threat to national aspiration. War rape and forceful impregnation had tactically aimed violence at women's bodies and reproductive capacities. Sexed living-being had been at the intensive crossroads of war, nationalism, patriarchy, and Cold War strategy.[31] The new state attempted to rehabilitate raped women as war heroines. Nonetheless, many women found their experiences silenced and their lives devastated.[32] Doctors and abortionists associated with International Planned Parenthood joined Bangladeshi doctors to provide services for the estimated twenty-five thousand women who had become pregnant from rape.[33] On the heels of this horrific sexual violence, economization arrived to intensify the necropolitical stakes of reproductive and sexed life: "No civilized measure would be too drastic to keep the population of Bangladesh on the smaller side of 15 crore [150 million rupees] for sheer ecological viability of the nation."[34] Further, the nation must be mobilized "at all levels" with a "total commitment" to its solution.[35] Too much life must not be born in the name of forging better collective economic futures. Family planning infrastructures spread exuberantly in the name of managing excess life.

While initially the new state aspired to strike an independent course of socialism divergent from U.S. alignments, it was not long before an Open Door Policy invited U.S. investment. U.S. foreign policy, in turn, had prioritized Bangladesh as a globally critical site of overpopulation that could threaten Cold War strategies.[36] The United States' own report on the war highlighted the ongoing crisis of "population pressure."[37] An order of state and nongovernmental activities was swiftly set in place by mutating emergency relief projects but also by reinvigorating the network of family planning schemes that had already been put in place before the war. The World Bank offered a loan, and USAID turned on the spigot of family planning funding, with experts from Johns Hopkins and Harvard following the flow. Paul Demeny (a collaborator with Notestein and Coale and prominent

American population studies scholar) described the effusion of infrastructures unrolling in Bangladesh in the early 1970s: "the program as is now planned will employ a small army — several tens of thousands of workers. Besides the personnel — administrators, doctors, nurses, midwives, lady welfare visitors, supervisors, female family welfare workers, drivers, peons, and so on — there will be the bewildering paraphernalia of things — from buildings to flip charts, from battery-operated slide projectors to four-wheel-drive vehicles, from motor launches to surgical equipment and supplies of pills, IUDs, injectables, and all the rest. By any measure, the plan envisages a formidable organization, delivering a sophisticated technology."[38] Foreign experts and funds spurred this exuberant unrolling of experiment, yet it was largely accomplished by Bangladeshi experts, social scientists, organizers, administrators, doctors, and field staff.

While Henry Kissinger infamously dismissed Bangladesh as an irredeemable "basket case" (and notoriously supported, politically and materially, Pakistan's military attacks on Bangladesh in a perverse playing out of Cold War strategy), history has shown that the landscape of human-achieved destruction in Bangladesh had set the stage for the creative explosion of Bangladeshi trials, pilot projects, and experiments, generating innovative techniques of development that would become global models. Postwar and postcolonial precarity became the fertile ground for the flourishing of inventive techniques for attaching life to economy, for bringing to market, for individualizing and choreographing choice, for new forms of debt, for social enterprise, as well as for averting life in the name of speculative prosperity. Precarity fed experimentality, which promised to attach survival to a better economy.

Thus, from the 1970s through the 1990s, Bangladesh was a site of tremendous innovation in experimentality with global ramifications. "Bangladesh has been a virtual laboratory for research," opens a typical 2007 World Bank report.[39] Bangladesh was the birthplace of microfinance and the Grameen Bank, which now operates internationally supporting microcredit and "banking for the poor" projects for women in the Philippines and Morocco, Brooklyn and Toronto.[40] It is home to the world's largest NGO, BRAC (originally the Bangladesh Rehabilitation and Assistance Committee), which in 2015 had over 100,000 employees, mostly women. Nobel Peace Prize–winning Muhammad Yunus (founder of the Grameen Bank) and decorated Sir Fazle Hasan Abed (founder of BRAC) are among the most globally lauded superstars of development, with BRAC celebrated as "undoubtedly the largest and most variegated social experiment in the developing world."[41] Yunus describes Bangladesh as a "living laboratory," a site of "innovative social and business thinking" that helped to generate microcredit.[42] Bangladesh was the cradle of the largest and oldest social marketing organization, PSI (originally called Population Services International), which in 2015 operated in more than sixty-five countries.[43] These examples are just some of the largest and most globally extensive manifestations of Bangladesh's history of experimental exuberance.

In this exuberance, NGOization became the dominant form of governmentality in Bangladesh, giving rise to a set of administrative and bureaucratic forms now extremely common throughout the world: the pilot project, the endless production of reports, the creation of a social class of experts, and the incessant proliferation of experiments.[44] NGOizaton does not merely describe a form of nonstate governance and service but names how "issues of collective concerns are transformed into isolated projects."[45] NGOization produced something like what Frantz Fanon called a "splintering" occupation. Instead of a centralized disciplinary rule, NGOized experimentality offers an ad hoc, continuously refreshing, transnationally attached, locally organized patchwork of unevenly distributing funds, technologies, practices, infrastructures, experts, services, collectivities, and workers from thana to thana, and from village to village.[46] Communities organized themselves in forms legible to NGOs as a route to becoming of interest to governance, to access care and funding, to make ethical publics, or to roll the dice on life chances.[47] In turn, foreign public health experts, doctors, feminists, community organizers, sociologists, anthropologists, family planning specialists, and humanitarian researchers were among the bestiary of experts from around the world who came to work with Bangladeshi NGOs: to test, to study, to train, to develop mobile protocols and best practices.

What this book's story underlines is how this era of experimentality in Bangladesh invented *neoliberal* techniques for the economization of life. Many scholars have shown how neoliberal techniques are characterized by the creation of audits, securitizations, surveillance, inclusions, and exceptions in the name of giving capitalism free run, of opening life to capital,

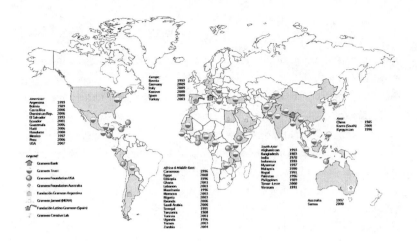

BRAC ACROSS THE WORLD

while tactically ignoring the externalities of injury created in the process.[48] Hence, neoliberal governmentality is not a mere retraction or deregulation of activity but a highly selective rearrangement of the terms, attentions, and *inattentions* of governance. Experiment too is a selective practice of attention and inattention. Most family planning experiments were akin to field experiments, epidemiology, or marketing research. When conducting sociotechnical experiments in "society," the controlled conditions of a laboratory are not possible. Field experiments take the world as it is, and seek to merely insert a single intervention and see what happens, without trying to control the environment of the experiment itself. The general milieu of precarity is a given, and the specific intervention—a contraception, an educational campaign, a drug, a loan—is introduced to see if it might induce a shift in the observed matrix of relations: will girls stay in school longer, will women have fewer babies, will fewer babies die, will more women work? Mass sterilization projects totaled acceptors but not side effects and mortality.[49] Evidence is highly specific, thus narrowing the set of claims that can be made from it.

Moreover, family planning experimentality in its "best practices" form phantasized a presumed liberal subject who freely chose to participate, who could be motivated and incentivized to accept interventions.[50] You were invited to be an experimental subject, and yet you were left on your own otherwise. The ubiquity of family planning experimentality targeted at stimulating an individual "choice" became a kind of neoliberal infrastructure. The interventions were "appropriate" to the setting: cheap, individualizing, in the form of a commodity, not requiring a working state, safe environs, or primary health care. Experimental intervention often aimed at

---

7.1–7.3 (OPPOSITE) The global spread of practices developed in 1970s Bangladesh. Figure 7.1 (top) shows the Grameen Foundation's global reach as the spread of microfinance from Bangladesh under the Grameen brand (Grameen Creative Lab, 2015). Figure 7.2 (middle) advertises the transnational spread of BRAC's operations from its start in Bangladesh, such that BRAC is rendered here as a global organization (BRAC Annual Report, 2013). Figure 7.3 (bottom) shows the global reach of PSI, which emerged out of the Family Planning Social Marketing Project of Bangladesh in the early 1970s and is now a transnational organization, the largest of its kind. (PSI Progress Report, 2013)

individualized and minimized cost-effective technological fixes that have as their most durable outcome the reproduction of an infrastructure of experiment.

In other words, the selective and individualizing structure of experimentality was particularly generative of neoliberal technique. This inventiveness was generated through the figure of population, through the project of family planning, in the name of women's choice, through designations of excess, and on the flesh of sexed living-being. It was facilitated at its foundation by disaster, violence, and an environment of installing capitalist relations. Experimentality presumes a world open to intervention and productive of change, but even more than this, neoliberal experimentality was premised on a sense of the world as generative not despite of but *because of* its precarity.

In these ways, Bangladesh was an important global node in the invention of postcolonial neoliberal practice. This inventiveness was *postcolonial* in that it was not conditioned by a corresponding effort to dismantle an already existent welfare state. Instead, experimentality flourished because of the very fragility of the new state, its lack of infrastructure in the aftermath of disaster and colonial pasts that produced conditions of profound precarity despite its profuse biodiversity and cultural abundance. While Bangladesh is often figured in dominant development discourses as a site of naturalized poverty since time immemorial, precarity in Bangladesh is fundamentality a historical achievement of the last century. Moreover, from 1970 to today, the shape of precarity has undergone dramatic change. Bangladesh became a site of cheap and disposable female labor in international supply chains of the garment industry. Rapid urbanization has made Dhaka one of the planet's megacities regularly vying for the title of least livable city. Unregulated industrialization has turned rivers black with pollution. Agriculture has been industrialized and monoculture crops prevalent. High-yield rice monocrops and debt for seed has transformed rural life. Millions of men and women have left Bangladesh in transnational flows of migrant labor. Bangladesh is one of the most affected regions of intensified climate change. Through the spread of microcredit, new relations of debt have become commonplace. Moreover, NGOization and experimentality tethered Bangladesh experts in new ways to transnational circuits of accumulation, debt, and research. Precarity was and is a recomposing accomplishment.

In turn, precarity has been generative of some kinds of infrastructures and not others.

No matter the result of the experiment, no matter if results are achieved as intended or if no change is measured (as often happened), what is reproduced is the infrastructure of experiment itself. The numbers, the protocols and best practices, and the reports would then work their way up the transnational NGO food chain: the grassroots organization strategically legitimizes its work with a report to its funder, a larger NGO, who passes it up to the donor organization, who celebrates its efficacy in another colorful graph at conferences circulating in an economy of numerical legitimacy. The economy of numbers calls for more intervention. If the experiment does not work, a better experiment is needed. What was built out of the figure of "population"? A refreshing infrastructure of experiment.

Family planning was far from the only arena in which reproduction became experimentalized in the late twentieth century.[51] This moment saw the invention of new biomedical technologies such as in-vitro fertilization, surrogacy, and prenatal diagnosis aimed at assisting fertility.[52] Accelerating since the 1950s, a tremendous medicalization of birth and fertility occurred. And more than this, the very generative and recombinatory capacities of genes, cells, oocytes, and metabolism have been harnessed toward a biotechnology revolution.[53] New arrangements for outsourcing clinical trials created an international division of experimental labor.[54] In contrast, subjection to experiment in family planning is typically denied the status of labor and instead designated as a kind of assistance or gift, even if it seeks to create new drugs or devices, or if incentives are used.[55] For example, the globalized experiments with Norplant, Depo-Provera, and quinacrine in the 1970s and 1980s were hidden drug trials legitimated as a form of assistance.[56] The experiment was purportedly ethical because, in a context of population crisis, benefits (preventing birth) purportedly outweighed risks (side effects with no follow-up care). Moments of attaining consent were also "motivations," caught within a global web of unequal constraints and incitements.

In the 1980s, feminists in Bangladesh exposed and denounced this economy of unregulated contraceptive and pharmaceutical testing.[57] The feminist Declaration of Comilla of 1989 went even further, denouncing the experimental reengineering of life writ large across the entangled domains of agriculture, development, genetics, and population control as "aggravating the deteriorating position of women in society and intensifying the existing differences among people in terms of race, class, caste, sex, and religion."[58] On the one hand, in the 1970s and 1980s, there was a rising denunciation in Bangladesh and the United States of the use of women and children as guinea pigs of the global contraceptive industry. On the other hand, the critiques and exposures did not seem to deter the generativity of the research infrastructure itself.

This generativity is exemplified by the story of Family Health International. In 1971 USAID began funding the North Carolina–headquartered International Fertility Research Program (IFRP) to rapidly test new family planning techniques — such as menstrual extraction, quinacrine, and Depo-Provera — in Bangladesh and elsewhere, for worldwide use. The research infrastructure of IFRP evolved in 1982 into the transnational NGO Family Health International (FHI), which oversaw USAID-funded "multinational contraceptive clinical research" and other development research in over seventy countries, including more than one hundred projects in Bangladesh alone.[59] By 1987, FHI was awarded the first USAID HIV prevention research contract, which began the reorientation of its research infrastructures toward HIV. By 2014, FHI, now FHI 360, branded itself as a full-service multinational clinical research organization overseeing more than half a billion dollars of U.S. government research funds, only 1 percent of which were now for reproductive health (FHI 360 2013). Moreover, FHI drew on its infrastructure to spin off for-profit clinical research organization enterprises such as Clinical Research International and PharmaLink. The USAID-funded family planning experiments of the early 1970s were thus reproduced as the global clinical trial industry of the twenty-first century.[60]

What has been reproduced in the name of reproduction? What kinds of infrastructures have been built out of the project of reducing future human birth rates? What endures and persists, and what struggles and is extinguished? Biomedical and biotechnological research fundamentally depends on the living and dying capacities of humans, mice, fruit flies, and bacteria. It is crucial to understand that reproductive assemblies do not merely create and animate life. In the lab, the organism must often die in order to extract the generative difference the scientist is after.[61] Mice are autopsied, unfertilized oocytes are tossed out, and so on. Charis Thompson has described a biotech mode of life in reproductive science, characterized by designations of waste and disposability in the lab in the name of targeting particular kinds of animacies and forms of life.[62] In a similar way, family planning experimentality describes a process of extensive intergenerational extinguishments that are not merely about making life. It has been as much constituted through uneven distributions of chances, of side effects, of losses, of dangers, and of designations of waste and surplus as part of the range of "responses" to projects that "assist" life. Experimentality, as an extensive assemblage, can enter a threshold of indistinction with capitalism's own attachments to accumulation, inclusion, and change on the one side, and externalities, loss, and waste on the other. According to a capitalist experimentality, the world is open to perpetual rearrangements and losses, where new conditions for creating value can always be generated even, perhaps even especially, out of the very precarity that is its externalized side effect.

In the case of population control, the future generations of the precarious are explicitly designated a kind of surplus life: sacrificeable and unvaluable to the economy, thus better never to have been born. Such designations went hand in hand with precarity: the more precarious life was, the more life had already been open to violence and struggle, the more it became attached to calculations of surplus life. The harder it was to get by, the more future generations could be calculated as excess. Moreover, tallies of future lives better unborn implicated and devalued already existing lives. Surplus populations, from the perspective of capitalism, are already disposable lives, whose future continuance need not be invested in. Participating in family planning projects, thus, could be fraught with a double relation: an opportunity to tactically care for one's body that also reinscribed one's relation to surplus life. This relation was material: to enter an experiment was to trade the risks of an intervention for the risks of exclusion or doing nothing. As family planning became increasingly tethered to child immunization in the 1980s, entering into the experimental relation combined a shift in an infant's chance of not dying with the prevention of being born at all through technologies that themselves bore risks to mothers' health. Experimentality thus produced infrastructures that promised to rearrange risks, and did so materially in ways that went beyond what was calculated in experimental protocols.

While designations of surplus life in the late twentieth century were carved out of distributions of precarity and abandonment, they were also mobilized to generate yet other forms of productivity. An export processing zone (EPZ), for example, takes advantage of spatialized deregulation, unemployment, and gender oppression to create reservoirs of cheap, disposable labor, as Melissa Wright (2006) has documented. Bangladesh currently has twenty-one special economic zones and the government promises to open one hundred more.[63] The success of Bangladesh's EPZs is explicitly analyzed by business discourses as relying on the region's abundance of "surplus labor." The 2012 Tazreen fire that killed 117 garment workers and the 2013 Rana Plaza building collapse that killed 1,129 people further attest to the disposable status of Bangladeshi women workers from the profit-maximizing viewpoint of global supply chains. The enormous infrastructures of family planning, development, and global health were built out of intervening in such death and precarity. At the same time, they were erected out of precarious life as one of their raw materials.

What is saving a life worth? What is the price for a life not to be born? The history of differential life worth, and the ways even well-intentioned research became caught in its regime of valuation, is exemplified in Bangladesh's oldest and largest population laboratory, which has hosted over five decades of continuous experiment. This world-famous field site is located in Matlab thana, a rural area not so far from Dhaka (initially an hour by boat; today four hours by car), and one of eight major field sites, including two in the Dhaka slums of Kamalapur and Mirpur, all operated by the International Centre for Diarrheal Disease Research, Bangladesh (ICDDR,B), headquartered in Dhaka.[1] The "surveillance" projects overseen by ICDDR,B extend to every one of Bangladesh's sixty-four districts. The ICDDR,B and its Matlab field site joins the Grameen Bank and BRAC as another globally important expression of experimental exuberance that arose in Bangladesh.

Matlab is a watery land formed amid the delta of the Meghna River, which joins the great Ganges Delta and then works its way to the Bay of Bengal. The regions around the Bay of Bengal have long been considered a site of endemic cholera, where the bacteria *Vibrio cholerae* flourishes on the copepods that make brackish water their home. The Matlab field site was first established in 1960 as the Cholera Research Laboratory (CRL), a Southeast Asia Treaty Organization (SEATO) Cold War military research project between the United States and Pakistan to conduct cholera vaccine trials.[2] There was already a census of the area from a smallpox eradication campaign and the household census cards were still available and easily updated: even numbers for the vaccinated, odd numbers for the control group.

By 1963, an infrastructure for cholera vaccine field testing had been assembled, which included a barge hospital that could operate as a floating field lab and assist urgently ill cholera patients who were provided with emergency care as an implicit contract for participation in the field trial. At the barge, patients were brought back to life with intravenous rehydration. There was a transportation system of motor boats that could both bring researchers from Dhaka to the field site and transfer sick people to the barge. A large cohort of Bangladeshi field-workers, initially local men (but later almost entirely women) trained as administrators of the vaccine and as dedicated data collectors who worked seven days of week, traveled from house to house in their assigned area, sleeping where they could, visiting vaccine recipients daily, and in the case of diarrhea taking daily rectal swabs to be delivered to the main lab. The impressive infrastructure included vaccine injector guns for mass-vaccination campaigns; standardized forms to record vaccine administration and track cholera prevalence; a base laboratory where stool samples could be examined; a trained crew of women "data recorders" to whom forms were submitted and tabulated; medical officers from the United States who served as directors; doctors and aspiring researchers from Bangladesh who manned the barge, oversaw the field-workers, and did the day-to-day work of research and health care. A Bangladeshi anthropologist was incorporated into the work of the site. While U.S. scientists tended to be globe-trotting public health experts from Johns Hopkins or Harvard, the majority of staff were Bangladeshi scientists, researchers, field-workers, and support personnel. Historically, ICDDR,B is the most important public health and biomedical research institute in the country through which Bangladeshi researchers have been able to launch successful scientific careers with transnationally circulating research.

The ICDDR,B is animated by the efforts and personal commitments of the Bangladeshi scientists and staff that have built this rare internationally recognized research institution in service of local people. One senior scientist I spoke to, who began his career manning the Matlab barge in the 1960s, had lost his mother and uncle to cholera as a child, but now IV treatment could miraculously revive people previously resigned to death. Another senior scientist sought to design projects that would leave tangible benefits in the form of concrete services for Matlab rather than just extract data. In the 1970s, both U.S. and Bangladeshi Matlab scientists had done relief efforts after the Bhola Cyclone and during the famine, and the work at Matlab often carries this spirit of service forward. Bangladeshi researchers risked their lives keeping the Matlab hospital open during the war. The ICDDR,B and the Matlab field site are an impressive achievement of fashioning a globally important site of postcolonial science in a resource-poor setting.[3]

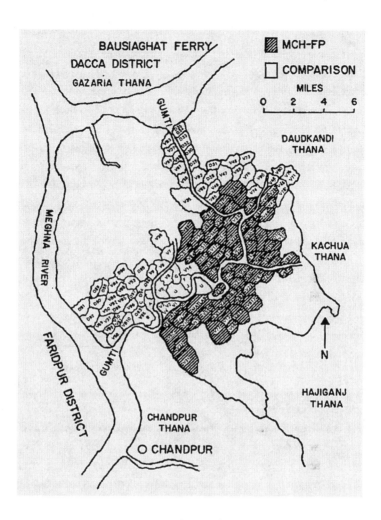

8.1   The Matlab field site in 1978. The map shows how the experimental region
for demographic surveillance was divided between intervention and control
areas. This split was made following the Contraceptive Distribution Studies
of 1975–77, and toward its elaboration as a Maternal Child Health and Family
Planning (MCH-FP) study in 1978. In 1978 some 174,500 people lived in this experimental region, and the surveillance infrastructure switched from field
worker visits every two to three days to weekly visits. (Aziz and Mosley, "The
History, Methodology and Main Findings of the Matlab Project in Bangladehs,"
1991)

The rare multidisciplinary focus on the perniciousness of diarrhea (unfashionable in global research) was exemplary of ICDDR,B's commitment. Bangladeshi ethnographers and sociologists worked together with physicians, epidemiologists, biochemists, and microbiologists. At the same time, Bangladeshi researchers collaborated with prestigious colleagues from the United States, who helped local scientists launch their own global careers. Young research assistants from the 1960s rose to become principal investigators on large grants from global funding agencies. Moreover, researchers at Matlab were known for their dedication, long hours, and the quality of their data that gave rise to an improbable globally famous research institution in a remote corner of the world. In the ICDDR,B archive, research counted "deaths averted." The language of averted births cannot be found in their records. The Matlab field station, as well as the ICDDR,B headquarters in Dhaka, both have cholera hospitals, where their treatments save the lives of nearly all who pass through their doors. The research infrastructure at Matlab pulled together manifold and sometimes conflicting motivations and lives, generative mixtures of sacrifice, nationalism, and extraction.

Also critical to the infrastructures at Matlab was the reenrollment of the people living there into repeated vaccine trials. These trials were remarkably unsuccessful. The promise of a vaccine that could rid poor rural communities of the perils of cholera was, and is, unfulfilled. Counterintuitively, the lack of positive findings in each iteration of a vaccine trial called forth yet another larger, more elaborate experiment. In 1963, 14,000 people in twenty-three villages. In 1964–65, thirty-five villages different from before. In 1966–67, with 40,000 children, in 1968–69 with 45,711 other children. In 1974, with 92,838 women and children, and in 1985, 183,826 people.[4] What was built from these repetitions was not a locally usable vaccine but instead a site and scale of research: a dense network of people and relationships. While a vaccine for cholera would be a life-saving accomplishment for people around the world, what the repetitions of these failed trials produced was an infrastructure for repeated research, turning the population of an entire region into ongoing experimental subjects. Research quickly expanded from cholera to include demographic and health data. It became a rare "Third World" field site with the capacity to collect frequent, intensive observations of some quarter-million research subjects: daily from 1963 to 1970, every one to two days from 1972 to 1974, every two to three days from

8.2 A Matlab field worker's data collection equipment in 2013 as she visits a household on her regular route. A newer PalmPilot data entry system is accompanied by the traditional paper forms and record books. Field workers are hired from the region in which they collect data, and everyone here has been asked the same questions before. (Photograph by author, 2013)

1974 to 1978.[5] Thus, what initially began as a site for (Cold War–militarized) cholera vaccine field trials was by 1966 turned into a general "demographic surveillance site," contacting 212,328 people in 122 villages daily. In 1968 the scope of the demographic surveillance doubled to encompass 224 villages.[6] Moreover, it created a research infrastructure open to repurposing and repetition, in which acts of experiment became habitual and ordinary. Matlab scientists I spoke with described a relationship of trust and obligation, a patron–client relation that makes Matlab possible. At the same time, researchers recounted the many ways that field-workers must constantly persuade and compel Matlab's people into this ongoing experimental work. Matlab became an arrangement of health services, of rescue from the grasp of cholera death, and an offer of employment to local women and men in exchange for experimentalizing the region. Matlab also improved on or filled absent state functions. It built durable systems of immunization, safe birth, cholera treatment, and other health services. In so doing, Matlab became the longest-running surveillance site in the world, still in operation today, and hence among the densest cumulative sources of data on a "developing" country. Matlab points to the postcolonial precursor to big data.

The history of Matlab turned during the 1971 war. Refugee camps in India struggled with an acute cholera outbreak. Building on Matlab research overshadowed by the unsuccessful, resource-intensive chase for a cholera vaccine, desperate Calcutta doctors, working with limited saline IV stocks, began to administer a simple mixture of water, glucose, and salts that could be taken orally.[7] It was this oral rehydration therapy (ORT), not a vaccine, that became the celebrated invention of Matlab once the war was over. In field studies, it was shown to be a cheap alternative to intravenous rehydration, requiring little expertise to mix and minimal medical infrastructures for support.[8] Cholera did not have to be a fatal emergency in the same way. Over the 1980s, BRAC was mobilized to teach ORT to women, who were enjoined to use the widely available remedy at home to reduce infant mortality. The ORT packets would join birth control pills as U.S.-subsidized items to be widely dispersed to households in rural Bangladesh and promoted with social marketing. Births averted (for a price of US$65 a prevented pregnancy) was joined to preventing deaths (at 5 cents per packet).[9] The built ecology of cholera had shifted, but so too had a price and a form been put on the value of preserving life.

Put simply, ORT redistributed possibilities of not-dying. It is arguably the single most significant intervention to come out of Bangladesh's experimental exuberance. Coupled with infant immunization, the wide prevalence of ORT is credited with an impressive reduction of under-5 mortality measures in Bangladesh (from 239 out of 1,000 in 1970 to 41 in 2012).[10]

While oral rehydration is indeed a crucial life-saving technology, it nonetheless moved the prevention of death from diarrhea away from water and sanitation management to dispersments of inexpensive emergency medicine administered near the brink of death. It is celebrated as an example of an *appropriate technology*, a pragmatist technology suited to off-the-grid settings.[11] It could cheaply increase life expectancy without a related improvement in conditions. As an effective intervention, ORT is celebrated without having to insist upon the supports that could have been built instead. Contemporary cost-efficiency analyses of ORT suggest that its ubiquitous contribution to reducing mortality now reduces the cost-effectiveness calculation of infrastructural projects to provide clean water and sanitation.[12] Yet, more careful studies of ORT and mortality, including a ten-year study at Matlab, have shown no definitive reduction of mortality with home use of ORT.[13] While ORT administered by health care workers has certainly reduced global mortality from diarrhea, saving millions, the same cannot be said with confidence about its individualized use without supports.[14] Moreover, as a technology of not-dying, it does not do away with the experience of diarrhea. The ORT prevented mortality but not morbidity. Surviving childhood diarrhea is correlated with stunted growth, nutrient deficiency, and brain changes.[15] Death by cholera was now preventable by a cheap emergency measure that did not demand changed conditions to world orders, and in this way enacted the logic of contemporary emergency medical humanitarianism, but as a ubiquitous fix for a state of chronic precarity.[16] Thus, ORT averts death but cannot hope to undo the uneven distribution of diarrhea itself.

After the war, research at Matlab was no longer home to a SEATO–Pakistan project. It became a national population lab funded by USAID and the National Institutes of Health (NIH) and was renamed the Center for Health and Population Research. The research infrastructures of Matlab were reoriented toward the project of family planning and "contraceptive distribution," reflecting the priorities of the new state, and also of U.S. foreign policy. Hence, a new set of sexed futures, toward economic development understood to be tied to population control and preventing future births, recomposed Matlab research. Over the 1970s, Matlab became the most internationally influential "population laboratory" for developing family planning practices.[17] Projects of reducing future poverty by preventing births repurposed infrastructures built to reduce cholera deaths.

The biggest single family planning experiment at Matlab was the Contraceptive Distribution Project that was conducted from 1975 through 1980 (encompassing some 233 villages) and was particularly known for its door-to-door "inundation" approach.[18] House-level contraception provision and daily contact by local field-workers (now women) turned the intensive surveillance of cholera research into the "continuous motivation" of family planning. It was in this era of research that increasingly large sums of funding began to flow through the field site. Its infrastructure of household contraceptive disbursement combined with basic maternal and infant health care would become the model for a national family planning system. Yet no national system could hope to achieve the highly monitored and rigorous practices of the Matlab field site that were made possible by its dense research platform. Experimentality at Matlab was not merely a local formation but a transnationally attached assemblage built over decades. Experimentality is not just the protocol but the extensive social-technical ensemble without which results do not hold.

By 1978 the experimental site was nationalized as the International Centre for Diarrheal Disease Research, Bangladesh, a nationally supported and world-renowned research center with a headquarters based in Dhaka and its now famous Matlab field site. The ICDDR,B became the go-to site for researchers from the United States working on family planning, economic development, nutrition, and public health. Its experimental activities produced a local economy of employment, services for residents, and permanent posts for Bangladeshi researchers, while at the same time foreigners served as senior management or donors, as well as fly-in researchers and students who would come for training or to run experiments through the research platform. Not only doctors but also demographers, sociologists, anthropologists, family planning experts, NGOs, and feminists have converged on the site and its rearrangements of labor and information retrieval. Research findings in Matlab were transported by such foreign researchers geopolitically to circulate as universalizable and mobile findings not especially about Bangladesh at all but as generalized features of "underdevelopment." Grameen and BRAC both used the site to test programs; BRAC alone conducted thirty-one research projects at Matlab in the 1990s.[19] The site has gathered over half a century of postcolonial thick data. Attached to this exuberant experimental infrastructure remained the cholera hospital with its chronic daily work of bringing people back from the brink of death.

In experimental exuberance, life and death were the epiphenomena by which an infrastructure reproduced itself. In other words, in contrast to a more familiar nineteenth- and early twentieth-century biopolitical story of infrastructures produced to govern populations, this is a late twentieth-century and contemporary story where precarity becomes the raw material

8.3 A view of the data collection volumes of the demographic and health survey at Matlab that exemplifies the history of postcolonial thick data. Field-workers come here to add the data they have collected to the records. The volumes are in the process of being digitized, and the paper data entry forms of the past are being replaced by handheld computing devices. These volumes hold over four generations and fifty years of data. (Photograph by author, 2013)

for and the effect of the continued flourishing of some kinds of infrastructures and not others. Research infrastructures bloom, cholera is treated at the threshold of death, sanitation infrastructures are not built, births are averted, numbers accumulate.

At Matlab and elsewhere, the economization of life has been composed by the necropolitical trilogy of death, not dying, and not being born. Conditions and threats of death legitimate the recompositions of risk that experimentality as a mode performs. Specific experimental interventions accumulate global resources while the more general structural conditions that produce precarity are mostly left as is. This is not the fault of the researchers, nor their innovative work. It is rather to describe the structural conditions constraining what can be done. Life is both necessarily saved and structurally devalued when cheap and minimal interventions at the point of death become infrastructural, as in the case of cholera and ORT. These minimized investments into preventing death stand in a perverse juxtaposition to the exorbitant efforts to prevent future life, which took sexed life and women as their point of concern.

Geographer Ruth Gilmore, writing about American prisons as an industry built out of the racialized production of surplus life, defines racism as "the state-sanctioned or extralegal production and exploitation of group-differentiated vulnerability to premature death."[20] In other words, racism makes and manipulates the unequal distribution of life chances. Racism can now be articulated in this particular way because of the thick history of numbers and statistics that count life chances and death as part of their politics. In family planning, differential life chances became an explicit epistemic concern, written into the calculations of averted births and cost-effectiveness measures of ORT. Distributions of vulnerability were studied, measured, quantified, and intervened in. Vulnerability was the object of inquiry. It generated postcolonial thick data, exuberant infrastructures of experimentality.

Not dying and not being born had distinctly unequal price points. Yet these two figurations of the value of life were produced side by side. Family planning projects were the recipients of enormous flows of funds from the United States, while funds for projects to prevent diarrhea were scarce. Instead of building clean water and sanitation infrastructures, death from diarrhea is prevented by the emergency cheap fix. While scientists in the late twentieth century rarely spoke about human life worth based on "race" as a kind of biological difference, the experimental and calculative techniques of the economization of life were explicitly written into policies, budgets, and reports and published in papers. These practices were generative of new globalized racisms, turning differentially precarious life into the abundance of experiment and number.

Experiment does not have to work in this way. There are other ways of conjuring a politics of life. In learning about the history of population control and experimental exuberance in Bangladesh, I have long been inspired by the work of Farida Akhter, who is the director of an organization called UBINIG (Unnayan Bikalper Nitinirdharoni Gobeshona, or Policy Research for Development Alternative). Akhter was one of the leaders behind the Declaration of Comilla and was among the most fierce and prolific critics of family planning experiments in Bangladesh in the 1980s and 1990s.[1] She interviewed women who were experimental subjects at Matlab and unearthed secret Norplant trials.[2] The only feminist bookstore in Dhaka is run by UBINIG. Akhter is the author of many books, and has ongoing collaborations with international feminist scholars, such as Gayatri Spivak and Maria Mies. The press of UBINIG republishes Begum Roquiah's writings, and Akhter positions her own work in a lineage of Bengali Muslim feminists that threads through *Sultana's Dream* and beyond.[3] For me, she is an inspiring theorizer of reproduction who has influenced this project and others since I discovered her writing as a graduate student in the 1990s.

For more than two decades, Akhter and UBINIG (along with her collaborator, the poet, intellectual, and political activist Farhad Mazhar) have been participants in Nayakrishi Andolon (the new cultivation movement). Nayakrishi Andolon manifests the possibility of an alternative politics of life in Bangladesh that also harnesses experiment, and seeks to reactivate older practices of organic cultivation for the contemporary. It is connected to larger ecological and ecofeminist movements in South Asia.

While UBINIG explicitly rejects the categories and terms of development and demography, such as economy and population, it theorizes and promotes its practices toward ecology, regeneration, and pleasure using the Bengali language. Refusing the terms of population and GDP, of development and waste, is a potent political gesture in the larger context of ubiquitous NGOs. Though this refusal is important, Farida Akhter describes her

9.1 The seed storage hut at UBINIG's experimental farm in Tangail, Bangladesh. Each earthen jar contains the seeds of a different strain of rice or some other plant food. The jars are labeled with numbers and the yield results for each strain are recorded in a ledger. This seed saving and documenting work is done primarily by women. The seeds here are then freely available to farmers, who then return a portion of the seed back to the seed hut. (Photograph by author, 2013)

politics as regenerative: "creating the future from the present embedded in the past."[4]

Nayakrishi Andolon is a mixed-crop, seed-sharing organic farming movement organized around the "Shahaj way to Ananda," roughly translated as the simple body and spirit interconnecting way to joy of life.[5] In expressing "shahaj," UBINIG draws on the philosophy of Bengali poet and mystic Fakir Lalon Shah, born in the eighteenth century, who drew on and blended Sufi Islam, Jainism, Vaishnavism, and Buddhism to cultivate a practice of embodied life that radically rejects divisive categories of identity, caste, and hierarchy. His expression was through song, and exemplifies a Bengali Baul musical practice of rural philosophy. Drawing on this tradition,

9.2 Rice grains in newspaper packages and braided straw that farmers have brought to the UBINIG experimental farm. The inclusion of straw reflects their insistence on calculating the systemic yield of a plot, not just the grain of the rice, as well as the other attributes of the rice. Straw is not waste, as in monocrop farming, but attended to as part of the yield that can be used for a variety of purposes in organic farming. Long straw can capacitate varieties to thrive in flooding or other microecologies. (Photograph by author, 2013)

UBINIG nurtures song as a mode of collectively theorizing and cultivating a different kind of life.[6] Nayakrishi Andolon is envisioned as a mixture of old and new, local and translocal.[7]

With Nayakrishi Andolon, UBINIG fosters an alternative to debt for seeds, monocrop high-yield rice, Monsanto, and pesticide use. Nayakrishi Andolon is constantly doing research into plant varieties and their suitability for different microecologies. A central part of this work is the collection, testing, and sharing of seed, especially rice varieties, of which there is incredible biodiversity. At its experimental farm, seed savers lay out a patchwork of multicolored circles of seeds to dry in the sun to be stored in earthen jars, a daily labor of caring for biodiversity. In its experimental fields, UBINIG insists on calculating yield differently than in conventional agriculture, explicitly valuing what commercial agriculture considers externalities and waste: soil health, noncultivated plants in marginal spaces, and the straw yield — not just the grain of rice — count toward yield. According to Akhter, "Use of pesticide or killing life forms are essentially unethical and against the culture values of communities. Therefore, it is absurd to disregard ethics and morality for the yield of a single crop by killing lives."[8] What UBINIG calls systemic yield includes the whole plant, as well as the health of the soil, and the other organisms that coexist with it on a plot. Nothing is designated as waste. Noncultivated plants, it argues, are an important source of food for the landless poor.[9] Rice varieties with long straw that withstand frequent flooding offer hope for a future of intensified climate change.

When I was interviewing Farida Akhter, I gushed over how inspiring her work was for the way it put together two domains of reproduction: reproduction in ecology and reproductive health politics. She corrected me. Are they two at all? She has not been putting these two domains together. Human and nonhuman reproduction were already together. It was the logics of population and economy that pulled reproduction apart. To help me see this, Farida told me the story that a midwife told her. When asked how many babies she had delivered in her long career, the midwife did not stop at the people but continued with counting the goats, the rabbits, and the chickens. Sustaining life in time was a distributed relation among many beings. Reproduction was already happening in a collectivity of beyond-human relations. This is a simple and intense realization.[10]

Population control and the green revolution were simultaneous projects in Bangladesh, and other parts of the world, reordering human and plant life together for the sake of economy and against future crisis, separating conceptions of birth from sustenance. The epistemologies and infrastructure of the economization of life have successfully and forcefully designated human reproduction and the activities of regenerating ecologies as separate domains. I realized that my own critical imaginaries had been occupied by the terms of economy and population. I had accepted that reproduction happens in the body, and that economy is out there somewhere. In contrast, holding reproduction together, the politics of life and experiment can be done differently.

Nayakrishi Andolon does not claim to offer a universal solution to the ways population and economy have reordered the world. But it does provoke the question of what other kinds of relations and concepts are possible if one refuses "population" and "economy" as orders. If one rejects the logics of population, the liberal ethics of individuation, and the horizon of economy, how does one experiment with a different politics of reproduction?

ARC III | Investable Life

In 1992 economist Lawrence Summers gave a keynote lecture, "Investing in ALL the People," at the Eighth Annual General Meeting of the Pakistan Society of Development Economists in Islamabad. In his talk, Summers famously argued for the economic benefits of educating girls. At this moment, Summers was not yet director of Obama's National Economic Council, nor was he president of Harvard, but instead he was vice president and chief economist of the World Bank. Drawing on the postcolonial treasure trove of published data about girls and their fertility, Summers calculated that each year of schooling reduces a girl's future fertility rate by 5 to 10 percent, such that US$30,000 spent on educating one thousand girls would prevent five hundred births. In contrast, a typical family planning program that directly distributed contraception costing $65 to "prevent" one birth would accomplish the same overall reduction of five hundred births for the larger amount of about $33,000. Thus, he concluded that education was $3,000 cheaper than contraception for averting birth.[1] Education was as cost-effective as birth control. With Summers's calculation, a new phantasmagram was congealing around girls, fertility, and economic futures. Educating girls had been plucked out as the crucial variable for increasing the value of life.

What is the value of a girl? At the end of the twentieth century, it was now possible to mine the postcolonial thick data accumulated in the decades of experimental exuberance about family planning to produce narratives layering many different correlations: the cost of educating girls was correlated with a cost-efficient lowering of fertility but also with lower mortality and higher future income. For Summers, "educating girls quite possibly yields a higher rate of return than any other investment available in the developing world."[2] The purported returns on investment in a girl's education to GDP were twofold: educated girls were expected to earn more as future wage laborers, and their reduced future fertility would lower population growth, thereby raising GDP per capita. In the 1990s, then, a new tabulation of the economization of life was assembling around the figure of the poor non-Western girl. Fertility reduction had become so thoroughly associated with increased national economic productivity that more distal stimulants to averted birth, like education, could now be targeted (and mined for cost efficiencies) to amplify the results of fertility reduction for GDP. More than that, this calculus suggested that projects to prevent future births were best temporally pushed backward in the human life cycle to the pre-childbearing years of life. Preventing birth had become preemptive.

By the 1990s, the thick data inherited from Cold War/postcolonial experimental exuberance offered a cornucopia of correlations about girls that could be harvested — links between lower fertility and an abundant field of other measures. For example, increase education, lower fertility, increase age of marriage; lower fertility, lower mortality, raise future wages. Such cascades of risk caught the attention of both economists and liberal feminists. Risks were increasingly imagined as chain reactions, where an intervention in one variable could set off changes in many others. Population's cloud of relations made legible by the dense dashboards of development offered up many interconnected anticipatory levers that programs might pull on. The biopolitical equation expanded from the if/then logic of *some must be born so that future others might live more prosperously* into a multitude of interconnected anticipatory possibilities: *some children must be invested in so that future others might not be born, so that rates of return increase, so that future adults are worth more, so others live more prosperously.*

The value of a girl, then, was caught in this cloud of anticipatory calculations. Anticipation names both a temporal orientation toward the future and an affective state, an excited forward-looking subjective condition of yearning, desire, aspiration, anxiety, or dread.[3] At both epistemic and affective levels, anticipation makes the future palpable in the present. Entangled with speculation, anticipation orients finance, enlivening its probabilistic and gambling urges with feeling. Anticipation is in the dizzying, abstract heights of colorful population charts, and at the same time a lived embodied state of expectation, aspiration, and anxiety. As an expression of neoliberal times, anticipation is nonetheless a contested orientation. Whose dreams for the future are legible; what aspirations for another world appear sensible? Both feminisms and finance are intensely engaged with speculative futures about girls that can take the form of a dread of diminishment or an excited premonition of a better life. At the turn of the twenty-first century, girls' futures existed in a dense constellation of anticipatory relations, imagined as cascades of probabilistic reactions that could be preemptively intervened in. With the arrival of "invest in a girl" calculations, the differential value of life was becoming increasingly financialized, oriented through both anticipation and preemption.

Summers's innovative argument of investment in girls repurposed an older notion of "human capital," a Nobel Prize–winning concept crafted in its neoliberal form in the 1960s by Theodore Schultz and most notably Gary Becker of the Chicago School of Economics. Human capital is typically defined as the knowledge, skills, values, and health embodied in people that make them economically productive. In one of his earliest works, Becker defined human capital as "activities that influence future real income through the imbedding of resources in people. . . . The many ways to invest include schooling, on-the-job training, medical care, vitamin consumption, and acquiring information about the economic system."[4] The notion of human capital is not the same as that of a human commodity — a person who can be owned by someone else, bought, and sold. Rather, in economics, "capital" names resources that are used in producing goods or services but are not themselves commodities for sale (such as a machine in a factory). Hence, the term *human capital* designates the embodied capacities of a person that can produce future economic benefits for that person, her employer, and even her national economy. Human capital hails thoughts, feelings, memories, skills, and flesh as assets. For example, paying for someone to become educated is an investment in their human capital correlated with future payoffs in terms of higher wages for them but also a more skillful worker for her employer. Thus, human capital renders embodiment as a site for investment. Bodies become a site for an anticipatory, future-oriented calculation of value.

Becker developed the concept of human capital largely through abstract and theoretical mathematical models (rather than empirical case studies) concerning fertility, the sexual division of labor, and family dynamics, research that formed what he called the "New Home Economics" as well as the Economic Theory of Fertility. For Becker, choices about fertility within families were fundamentally rational economic acts in which children were repositories for a family's capital. For Becker and Schultz, all human behavior was already forward-thinking, rational, and engaged in cost–benefit calculi, and thus all human behavior, including *within* families, was best understood in economic terms, regardless of culture or history. Crucially, the concept of human capital shifted the iconic economic subject from a worker or a consumer to an entrepreneur whose body contained an alterable set of assets and risks, a reknitting of *Homo economicus*. As Schultz explained in his 1979 Nobel Prize lecture, poor farmers and women were, "within their small, individual, allocative domain," all "fine-tuning entrepreneurs."[5] Even the child becomes an entrepreneur. Or, as Geeta Patel suggests, they become "risky subjects" compelled to calculate uncertain fates through speculation.[6]

Like Stephen Enke, Becker's work involved calculations of the differential value of children. Becker compared the rates of return on a child's value from investments such as education with the rates of return on the bare "cost of children" whose economic value was as a skill-less child laborer.[7] In this way, children become vessels of investment haunted by a condition of lack. Cold War economists working at RAND and for USAID quickly took up human capital in their work calculating the value of children and modeling parental demand in the 1960s and 1970s. Looking back at the 1974 classified report on the threat of worldwide population growth commissioned by Kissinger that helped to spur the U.S. commitment to funding family planning globally, one can see that analysts had already strategically identified investments into "minimal levels of education" for women and "improvements to the status of women" as stimulants to reducing fertility.[8] Instilled into the strategic priorities of averting birth as a Cold War project were already the calculative preoccupations that would give rise to twenty-first-century campaigns to "invest in a girl." However, unlike the tabulations of negative value that were attached to all "underdeveloped" life in the Cold War demographic transition model, with human capital low value is improvable in the future under the right risk conditions. Her returns are so high because her value begins so low.

Importantly, this refocus on a girl's human capital had moved the calculated point of development intervention from fertility itself to education, from distributing contraception to "women" to producing the conditions for higher rates of return on "girls," a change that has come to dominate development programs at the turn of the twenty-first century. The figure of the "Third World girl"—typically represented as South Asian or African, often Muslim—has become the iconic vessel of human capital. Thoroughly heterosexualized, her rates of return are dependent on her forecasted compliance with expectations to serve family, to adhere to heterosexual propriety, to study hard, to be optimistic, and hence her ability to be thoroughly "girled."

In the context of hegemonic liberal economic imaginaries drawn into relief at such sites as the World Economic Forum or the Clinton Global Initiative, the Girl as human capital became the darling of philanthrocapitalist ventures, especially in the wake of the 2007–8 global financial crash. Prominent here is a campaign on "the Girl Effect" by the Nike Foundation with its deep corporate investments into narratives of harnessing individual potentialities and "just do it" agencies. The Girl Effect promotes "the Girl" as a feminist solution to the "world's mess."[9] Following the investment into a girl's education—"put her in a school uniform"—is an avalanche of purported effects leading to the increased value of her life to her village, to women's rights, to national production, and finally to world salvation: "invest in a girl and she will do the rest." With award-winning stylized graphics and swelling music, the campaign offers an equation of cascading correlations: "Girl→School→Cow→$→Business→Clean H_2O→Social Change→Stronger Economy→Better World." Through this trickle-up chain, "an educated girl will stay healthy, save money, build a business, empower her community, lift her country, and save the world."[10] In this phantasmagram, financial investment by a donor propels a chain reaction determined by global capital. At the same time, the Girl Effect offers up a phantasmagram in which the weight of the world's economic futures rests on the risky subject of the abstract and universalized girl-child. Investment into the Girl becomes a way for national and economic futures to rejuvenate themselves.[11]

If the Girl is not invested in, another more apocalyptic result is prophesied. "The clock is ticking": if she is not invested in by twelve, she will be married by fourteen, pregnant by fifteen, after which she may have to sell her body, and contract HIV, leading to her premature death. The Girl is a ticking time bomb of risk. "Chance" is translated into only two possible paths: the unproductive life and the productive life. The "fact sheet" accompanying the Girl Effect campaign presents two possible futures—one

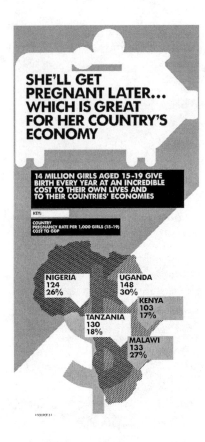

SHE'LL GET PREGNANT LATER... WHICH IS GREAT FOR HER COUNTRY'S ECONOMY

14 MILLION GIRLS AGED 15-19 GIVE BIRTH EVERY YEAR AT AN INCREDIBLE COST TO THEIR OWN LIVES AND TO THEIR COUNTRIES' ECONOMIES

KEY:

COUNTRY
PREGNANCY RATE PER 1,000 GIRLS (15-19)
COST TO GDP

NIGERIA
124
26%

UGANDA
148
30%

KENYA
103
17%

TANZANIA
130
18%

MALAWI
133
27%

› SOURCE 3.1

10.1   An infographic of the "world-changing ripple effect" that can add billions to world GDP. The Girl Effect is repeatedly attached to improvements in GDP, reanimating the economization of life as a preemptive investment into the Girl as a kind of human capital with high rates of return based on investing in her primary education. Her future financial returns come from her increased future income, reduction of future fertility, and the reduction of her costs to the state. (Girl Effect, "Economically Empowered Girls Can Stop Poverty before It Starts," 2013)

in which she is "invested" in and the other in which she is not—each accompanied by a chain of correlations, pronouncements of what she will become, offered not as probabilities (she might, she could) but as prophecy (she will).

In Girl campaigns, a new mode of visual representation replaces photographs of suffering and unmet need that decorated 1980s development and family planning literature. Instead, the Girl is namelessly ventriloquized as addressing potential donors in the first person: "Invest in me. It makes sense. Start when I am very young. Watch your investment grow, as I do."[12] Representations of the Girl are visually optimistic, hopeful, and animated. The Girl is represented abstractly as a charged data point, an ebullient cartoon, or an animated icon. Through music, color, and animation, the icon of the Girl is excited with enthusiasm, hope, ambition, and responsiveness to

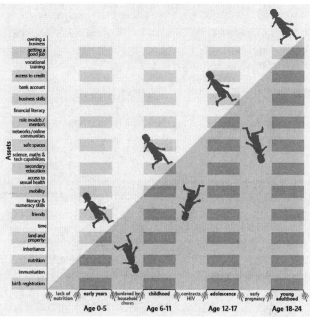

10.2 The invested Girl catches up to the right side of time, which then "impacts the world" as her clock transforms into a radiant globe. In this still from a Girl Effect campaign video, which premiered at the 2010 Clinton Global Initiative meeting, the Girl is an anticipatory figure on a lifecycle clock. The text declares, "We have a situation," inviting the viewer to feel anxious about the life outcomes of the Girl that are presented as in need of urgent interventions. If she is not invested in by twelve years of age, a negative cascade of risks will be unleashed. The invested Girl, in contrast, will be able to unleash her potential and conquer risks. (Girl Effect, "The Clock Is Ticking," 2010)

10.3 The Girl with only two possible futures, oriented by a cascade of either negative or positive outcomes, in a chart from Plan's "Because I Am a Girl" 2009 campaign. (Plan, "Girls in the Global Economy," 2009) transformation that joins Western liberal feminist imaginaries of empowerment to speculative finance. In donor-focused campaigns aimed at North American and European elites, the audience's hopes and anxieties are called upon to be cathected with the figure of the Girl; their individualized agencies can activate hers. *You* can change *her* life. "You can be the one."[13] The Girl Effect needs to be triggered by the attention, investment, and education of elites, even if she does all the heavy lifting.

The Girl is a phantasma, an affectively charged figure of a subject brought to life with numbers and animation. The Girl is a generic figure assembled out of quantification, speculation, and affect, a stereotyped abstraction of a subject "figured out" from a variegated patchwork of social science correlation and wishful speculation, of linked probabilities painted pink with tropes of agency imported from liberal feminism for a North American audience. The Girl is calculated as a risk pool that draws together a bloom of possibility, a bouquet of potential, a cluster of affect, applicable to any dispossessed condition anywhere, as long as it is "girled."

This visualization of the Girl as a generic figure charged with speculative futures repeats in campaigns by major transnational NGOs, such as Plan and its "Because I Am a Girl" campaign, and U.S. feminist NGOs such as Care, Vital Voices, the International Women's Health Coalition, and Women Deliver, as well as corporate responsibility projects about girls by Nokia, Chevron, Shell Oil, Exxon, Credit Suisse, Walmart, Intel, and Goldman Sachs. In the prominent "Because I Am a Girl" campaign of Plan, the Girl is animated as a colorful dot, a pulsing pie chart, a blooming flower, or a stop-motion living marionette all constituted in a vortex of probabilities that correlate girlness with either extremes of poverty and abjection or compliant and community beneficial forms of waged and unwaged labor. The world-changing effects of stimulating these correlations is represented as a pink and purple globe bursting with flowers that rise into the planetary stratosphere, offering an image of globalized economic growth as created through girled life.

Directed at feminist-positive potential donors, the "Because I Am a Girl" campaign also included a video for viral circulation in Canada through which the aspirational ethics of feminist-oriented donor girls are mobilized as they are invited to be "the one" who makes a difference because she is also a girl. In social media, donor girls are invited to insert their own

10.4 The potential of girls is thoroughly gendered in a still from Plan Canada's "Because I Am a Girl" campaign video (2009). Here the globe is covered with glowing data points that represent the potential of girls. As the potential of girls is unleashed, the globe of data points bursts into flowers and pink lights while an enthusiastic female voice explains, "It is proven that investing in just one girl will empower her family, improve her community, alleviate poverty, and will change the world." (Plan Canada, "Because I Am a Girl," 2009)

faces into circles, converting their own images into subject icons of "can do" data points for the campaign.[14] The corporate enthusiasm for promoting the Girl, and the enthusiasm of U.S., Canadian, and UK liberal feminist NGOs to couple with corporations to deliver Girl campaigns, is symptomatic of the affective bonds between Western liberal feminisms and financial logics. The Girl is imbued with the liberal feminist promise of individual agency, translated into a data point charged with the promise of value-added human capital. An emotional cocktail is thus mixed by the figure of the Girl: anxiety about global economic futures; empathy for the harsh and cruel lives of some girls; attachments to broken girls known or imagined; hope and admiration for the amazingness of girls; liberal feminist aspirations for agentic and equal life; personal desires to help and rescue; belief that technoscience will offer the right lever to be pulled, the right button to push, the correct commodity to purchase that will fix the world with-

10.5 The figure of the investible Girl. This image from 2009 promotes the UN–Nike supported Coalition for Adolescent Girls at the time of the Davos Economic Summit. Here, the Girl is described as both the poorest and the most powerful person. The image takes the figure of a population pyramid from demography and places the girl beneath it, where she is responsible for carrying the weight of population. This is just one example of the many ways the Girl is responsibilized with the future of herself, her family, and the world. (Coalition for Adolescent Girls, "The Most Powerful Person in the World," 2009)

out having to undo capitalism itself; and also darker fears of unruly youth, rioting and rebelling, dangerous and unproductive. The Girl is dense with contradictory feeling.

As an effect of non-innocent infrastructures of data collection and governmentality, the Girl is of interest to Goldman Sachs, Intel, and Nike precisely because the numbers designate her as a good investment with high rates of return, raising the question: what if the math had not added up, and in fact another object or life form was calculated as the best investment?[15] The popularity of the Girl raises the questions: What work does this phantasmagram do for capitalism? What is the Girl an alibi for?

Following on the heels of financial collapse in the United States and Europe, the enormous popularity of the "Girl Effect" was linked to the suggestion at the World Economic Forum of 2009 that the Girl offered a way of "develop[ing] a swift and coordinated policy response to the most serious global recession since the 1930s."[16] The Girl was on offer as an under-valued stock for future economic recovery, a cure to the "world's mess," a potent generator of increased GDP. As Plan's influential report *Girls Count* explains, "With adolescent girls the case is perfectly clear that the economic and human rights agendas are perfectly aligned."[17] This vision of alignment between capitalism and the Girl was buttressed by Goldman Sachs' own 2007 Global Economics Paper, "Women Hold Up Half the Sky," which cal-culated that "a one percentage point increase in female education raises the average level of GDP by 0.37ppt and raises annual GDP growth rates by 0.2ppt on average" because better female education is correlated to greater entry into the workforce. Moreover, "the combination of female education and pervasive wage discrimination can create a pool of well-skilled but in-expensive female labor. In itself, this may lead to more investment in indus-tries dominated by female labor. Higher returns to industries that fed on female-intensive light manufacturing played a role in the rapid growth rates seen in Southeast Asia."[18] According to Goldman Sachs, the investable girl is forecasted to perform well, with the usual market forecast disclaimers applying.

This abundance of investment into the Girl juxtaposes the hypervalua-tion of anticipatory "potential" with a set of unspoken devaluations: of births to be averted; of the less valuable women that uncapitalized girls grow up to be, a future form of life no longer worthy of investment; and of boys who offer lower rates of return. As the Girl is added up as posi-tive potential, the figure of the racialized young, possibly Muslim, male is anxiously constituted as her other: unruly, undisciplinable, and potentially dangerous. Further, *within* the figure of human capital itself resides the less discussed figure of expendable life, on sale for cheap to Nike's factories. Her rates of return are so high because her value begins so low, the "poorest of the poor." Thus, these anticipatory techniques that entwine data, visualiza-tion, finance, feminism, and economic development have gathered within the figure of the Girl designations of *investable lives, avertable lives,* and *ex-pendable lives* all at the same time.

If the Girl is haunted by devalued life in the form of the threat of non-compliant brown boys, the militarized implications are not coincidental. The U.S. Department of State, under Hillary Clinton, started its own fund for investment into women and girls, arguing that focusing on the well-being of women and girls promotes democracy, promotes stability, creates more opportunity in societies. That's just an absolute fact. . . .

If you look around the world at the areas that are unstable and are incubators of terrorism or other forms of violence, you will find women and girls being oppressed, being denied rights, being marginalized in a way that is dehumanizing . . . it is something that I see as Secretary of State that is absolutely integral to our approach to the kind of better and safer world that we are all trying to create.[19]

Reinvigorating old tropes that ethicize imperialism through the figure of the woman in need of rescue, the investable girl is always already a national security solution. This securitization of the Girl echoes a rare speech given by former First Lady Laura Bush that inaugurated the invasion of Afghanistan in 2001 with a public radio broadcast: "The fight against terrorism is also a fight for the rights and dignity of women."[20] Nicholas Kristof, a prominent proponent of liberal feminist forms of American development, offers up the Girl as a counterterrorist tactic, suggesting that "sometimes a girl with a book is more powerful than a drone in the sky."[21]

While the Girl was at first formulated in PR campaigns to mobilize donor, corporate, and policy audiences, the resulting spread of commitments to the ethic of investing in girls—which now ranges from the UN to the World Bank to the Gates Foundation to the U.S. Department of State—has increasingly prompted the practical question of creating programs to actually enact the interventions the Girl Effect dreams. "Put a girl in a school uniform" turns out not to be so straightforward. Getting to school can be difficult: there is labor to do, restrictions on mobility, or even larger problems like violent conflict or the aftermath of disaster. Educational infrastructures can also be a problem: no gender-specific bathrooms, not enough teachers or supplies, and paltry curricula, or an emphasis on rote learning. It turns out, not just any school will do. Girl projects are interested in investing in girls as individuals, not building public school systems. Moreover, it turns out there is actually very little detailed empirical social science research about how to design projects that educate girls in the ways the Girl Effect imagines. How, then, does one invest in the Girl?

While the Girl is a phantasmagram of data and anticipation, one of the most pervasive calls for practical action is to collect *more* data. One of the salient attributes of the Girl has been her risk of being inadequately registered in the infrastructures of data collection: she might not have a birth certificate, might not be registered in school, might not be engaging in formal labor. Even as she is thick with data she is contradictorily constituted as a remainder, neglected in the census, without proper registration, abandoned by Cold War development and patriarchal accounting. In response, the Nike Foundation and UN Foundation commissioned Maplecroft, a London-based corporate risk analysis firm, to conduct an analysis of the global risk landscape that governs the Girl, a project titled "Girl Discovered."

Maplecroft promotes itself as "the world's leading global risk analytics, research and strategic forecasting company" and specializes in making maps, indices, and data visualizations of "poorly understood" extrafinancial risks for corporations and other clients seeking to take advantage of risky areas of the world with potential high rates of return. It prides itself on its ability to plot data onto interactive maps that can zoom up to the planetary scale or drill down to a particular urban neighborhood. Maplecroft's data visualizations are not contained by the bottle of "the economy" or the aggregate of "population"; instead, Maplecroft promises to forecast at the finer resolution of subnational levels of risk that might hamper multinational corporate operations and supply chain logistics. Maplecroft's data visualizations do not idealize a world made up of national "economies" but instead emphasize the myriad multiscaled risks that shape a globalized tentacular supply chain capitalism. Here, the conjuring of capitalism's productivity shifts from GDP in the container of the nation to the profits made in planetary logistical arrangements, where production is materially stretched across space, across jurisdictions, and is thus no longer contained in a single factory or even country. The geographer Deborah Cowen has described the emergence of global logistics at the conjugation of military and business management practices as performing a "disarticulation of production into component parts that can be stretched out and rearranged in more complex configurations."[1] This is the stretched-out mode of border-crossing capitalism that Maplecroft visualizes and the Girl increasingly inhabits.

Clients can purchase a subscription to Maplecroft's rich portfolio of sector-specific risk analyses, such as the Human Rights Risk Atlas, the Climate Change Risk Atlas, the Terrorist Dashboard, and the Consumer Potential Index. In 2015 Maplecroft offered over 1,500 such indices and indicators to its clients.[2] Through this bevy of colorful indices and dashboards, it promotes a strategy in which taking advantage of "global growth opportunities" (in the "resource-risk" sites that used to be called "less developed countries") necessitates a new kind of "pro-active corporate social responsibility" in which companies are advised not just to engage in philanthropic charity that enhances the aura of their branding but also to integrate a more anticipatory practice in which corporations seek to mitigate or preempt extrafinancial risks that might benefit their supply chain operations.[3] For example, investing in AIDS-related health clinics for truck drivers would decrease labor inconsistencies and reduce risks to logistical operations for a delivery company operating in southern Africa.[4] Or when Nike invests in educating poor girls, reducing their risk of early marriage and fertility, the girls in turn become a more consistent global workforce. Thus, pro-active corporate social responsibility seeks to intervene in the "drivers" of particular risks within risk-saturated spaces of "growing opportunity," and thereby preempt blockages to supply chain logistics. Among the many indices, atlases, dashboards, and scorecards that Maplecroft supplies to corporate clients is the Women's and Girls' Rights Index, which measures, among other variables, how "companies operating in resource-risk countries" are susceptible to becoming complicit with rape by privately hired security forces, coloring sites of "extreme risk" red.[5] While the Women's and Girls' Rights Index is proprietary, Maplecroft, working with UNICEF, has made its Children's Rights and Business Atlas public.[6] In the optimistic phantasmagrams of the Maplecroft universe, each threat is always also an opening to prospect, as manifested in their phrase "global risk/opportunity."[7] Here, the surround of capitalism shifts and "the economy" as a container dissolves into a data-verse of planetarily dispersed multiscalar risk/opportunities twitching with data feeds.

This approach to risk forecasting also animates Maplecroft's work on the Girls Discovered project (renamed Girls Stats in 2016), which sought to gather and map data on girls from some two hundred indicators.[8] Ironically, one of the most touted findings of the Girls Discovered project was "an awkward truth. As a global community working in development, we are in the dark when it comes to girl data."[9] How could this be? While the argument for investing in girls was generated out of decades' worth of postcolonial thick data, the Maplecroft analysis suggests that this archive merely reveals a macrotendency toward a simplistic bundle of correlations that is already out of date. While the data are certainly thick, they are not globally comprehensive, nor are they in real time — at least not to Maplecroft's standards of fine-grained logistical risk mitigation. The archive of Cold War/postcolonial thick data tied to fostering national economies can only reveal so much about girls' risks in the world of global supply chain capitalism now.

Here, the comparison is not to other forms of development data but rather toward the horizon of *big data* for development and the phantasy of endless risk landscape analysis that Maplecroft sells. Big data's phantasy is particular, digital, streaming, and constant. As a start toward this new big data horizon, Girls Discovered offers an atlas of girl "vulnerability" that can drill down to the subnational level for India. Making its case for more data, the Girls Discovered project tracks data availability and "data gaps" where Canada, Australia, or European countries score particularly poorly as data generators, while Brazil, Swaziland, or the Philippines, sites of intensive-development NGO activity, score highly.[10] On data availability, Bangladesh is an exemplar of productivity. Thus, a new phantasmagram for the Girl was consolidating twenty-five years after the Cold War, aspiring to surround itself with real-time targeted data, where "data gaps are opportunities."[11]

If the cascade of correlations that makes up the Girl Effect are not so simple to spark as paying for girls to go to school, more data promises to make possible "unique, targeted programs" of intervention at the local level.[12] "Data saves lives," so the Nike Foundation and Maplecroft extol.[13] More data promises more "impactful programming" revealing the unique "drivers" of change for girls in different risk localities.[14] Striking the same note, Bill Gates, in his 2013 "Annual Letter" as head of the Bill and Melinda Gates Foundation, called for more data that can track feedback loops from

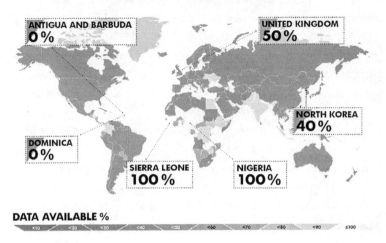

AGE AT FIRST BIRTH

ANTIGUA AND BARBUDA
0%

UNITED KINGDOM
50%

NORTH KOREA
40%

DOMINICA
0%

SIERRA LEONE
100%

NIGERIA
100%

DATA AVAILABLE %

<10   <20   <30   <40   <50   <60   <70   <80   <90   ≤100

11.1 Sample data visualization from the Girls Discovered project. Initially constructed out of the thick archive of postcolonial data about women and girls from family planning history, the Girl is now reconfigured as lacking data. Here, maps of data availability created by Maplecroft for the Girl Effect sort countries according to the availability of data on select variables, such as age at first birth, as in this map, but also healthcare, education, and access to income. In these maps, countries like Bangladesh and India, with long histories of development accounting, are data rich. (Girl Effect, "Global Availability of Data on Girls," 2013)

precise measures as keys to development's future. "So in contrast to several decades ago, where it was equally well meaning, now it is more like a business where you know the numbers, you see what you have to adjust, and that is letting us make huge progress."[15] Melinda Gates tied her call for big data to girls: disaggregating new flows of data by gender for crafting the UN's 2015 round of development goals.[16]

With the promise of streaming higher resolution counting, perhaps the numbers for investing in girls' education do not, in fact, add up? The 1990s call to invest in the Girl was originally formulated as a generic analysis, with the Girl as a generalized abstraction on whom risk converged along a universalized cascade of relations, applicable everywhere, and benefiting the phantasmagram of the macroeconomy. In contrast, calls for more Girl data open up the Pandora's box of asking for an always more refined cloud of correlation, for more precisely targeted and emergent "drivers" that can push or pull along the larger firmament of risky relations. Drivers are quite literally driven, by highly targeted adjustments that attempt to shift attuned measures on "dashboards" of risk. Girls are only one of the many risky opportunities that might be preemptively altered.

At the turn of the twenty-first century, the container of the economy, which is a view from a nation-state, can be eclipsed by the phantasmagram of a variegated globe pulsating with risk/opportunities miraculated through big data, a view from the perspective of the multinational corporation. How will the data be made to add up? Why should it be girls? What will be identified as investable life? Or expendable life? Or the next best investment in the developing world?

This newer vision of big data for development remains elusive. Big data, at its most promissory, calls for real-time numbers and real-time analysis of the "digital smoke signals" that global capitalism now habitually "exhausts."[17] The UN's Global Pulse project, launched in 2009, dreams a future in which streaming flows of data are donated by corporations as a kind of "digital philanthropy" and are harvested from mobile phones, which have rapidly spread across the globe, fostered not only by the spread of telecommunication markets but also by a previous moment's development excitement about mobile phones as a means to promote micro-entrepreneurship for women.[18] While the hope of big data requires a world exhausting information as its surround, the calls for big data perform concrete work in the present, reframing the Girl as not attuned enough, and the Girl Effect as needing more "targeted" drivers that might make investment more precise.

Cell phones are heralded as one of the most promising sources for real-time data on girls. More than data sources, cell phones are also increasingly imagined as a vehicle for giving cash "investments" directly to girls. If the "drivers" are so shifting and local, then why not put cash investments directly into the hands of girls themselves and responsiblize them for their own investments? "Financial literacy" and "access to financial assets" have become circulating measures of gender equality for the Girl. In 2013 the Girl Effect campaign began to focus on the "economic empowerment" of girls, where "we unleash the Girl Effect when we recognize that girls are economic actors who can speed up development in their community." Girls, therefore, need to be prepared with financial skills and even start-up financial assets.[19]

This turn to intensify the financialization of the Girl is multipronged. First, it has induced new metrics of financial inequality, such as counting girls as part of the "unbanked" or as bearers of "economic empowerment."[20] Second, educational projects have increasingly included teaching financial literacy—the "knowledge, skills and attitude to handle money well"—as a key human capital "asset."[21] Financial education is seen as more than just about pragmatic rationality; it is also "changing their aspirations and those of their community"—a reiteration of the older "calculus of conscious choice" of family planning.[22] Third, these cognitive human capital assets are joined by actual monetized assets for girls in projects that aim to directly give payments or e-bank accounts to girls. Projects of inciting girls with direct payments build on already established practices: such as giving financial incentives for participation in family planning (which began as early as the 1960s) and on more recent practices of giving stipends for families who agree to send their daughters to schools (as in Bangladesh), or broad-based income grants to the poor (as in South Africa).[23]

Mobile phones, then, do not merely exhaust data; they are vehicles for investing in a Girl's "financial assets" in the form of infrastructures of e-banking.[24] In this vision of the near future, programs will put cash in girls' hands through biometrically registered cell phone e-bank accounts, their cell phones will expel data for real-time analysis, and girls can be nudged through their phones toward making good choices.[25] Here, the epistemic infrastructures for making numbers are being transformed. Smart phones and the Internet, rather than surveys and field sites, become the new infrastructure for counting, marketing, and investing in girls. The Girl becomes an entrepreneurial subject who produces her own data.

From the perspective of the financial sector, girls are part of the planet's great "unbanked," the people who might yet be brought into the fold of institutional financial relations.[26] Girls are part of the "bottom of the economic pyramid," which "presents a significant growth area for banks in emerging markets."[27] Thus, the investment into the Girl is also a project of enrolling girls into particular kinds of economic activity beyond labor.

11.2 A Girl in Intel's 10x10 campaign video. Smiling girls from diverse non-Western locations each holding a white sign with writing edited in later during production completing the sentence "I am . . . ," such as "I am capable," "I am a good investment," and "I am an emerging market." This is a familiar representational strategy in Girl campaigns, where narrations ventriloquize a request in the first person to treat girls as sites of financial speculation. The girls holding the white board have no knowledge of the message they will be used to advertise. (Intel, *Not a Number*, 2013)

Ananya Roy, discussing the case of microcredit, describes girls as a kind of "poverty capital" for financial investment.[28] The Gates Foundation identified "mobile money" for the poor as a best "bet" of 2015, pointing to Kenya as the world's exemplar.[29] In Bangladesh, BRAC has already unrolled bKash, "one of the most exciting ecosystems for digital financial services" that allows financial inclusion through mobile phones to reach "where roads, electricity, and other infrastructure are still woefully underdeveloped."[30] With its extensive infrastructure of adolescent development program clubs for girls, BRAC has collaborated with a local mobile phone provider, Robi, to accelerate girls' cell phone usage, even developing a new app, called Maya (the same name as the old USAID-funded Bangladeshi birth control pill), especially for women and girls.[31]

The Girl is phantasized as a potential awaiting financial inclusion, or as

Intel's 10x10 campaign presents the ventriloquized girl: "I am an emerging market."[32] The Girl exemplifies the way finance capitalism creates value out of life, rendering life as something that either accrues or diminishes in value depending on how the riskiness of the milieu is gamed.[33] In the galaxy of alterable forces that data conjure, the Girl is an "effect" waiting to be "unleashed," just like other "growth/risk opportunities" for capital.

As I walk past the library en route to the subway in my downtown Toronto neighborhood, I often pass young women and men wearing blue canvas "Because I Am a Girl" vests who are raising money for Plan. While I usually just brush by, on a few rare occasions when I am a little less busy I have stopped and talked to them. I ask them about their training, and what they think about the Girl Effect, soliciting their opinions on the projects that they have learned about. One young man was particularly enchanted with the story of a project in Peru where girls started businesses raising guinea pigs. Working precariously as a minimum-wage canvasser in a North American context of chronic youth unemployment, raising adorable guinea pigs in a faraway place has allure. In turn, I voice a synopsis of my critique of the human capital approach to liberal feminism. I end up feeling deeply ambivalent. The young fundraisers are kind and enthusiastic while working for cheap. And my critique has a deflating effect. They too are young and precarious, even if at a different intensity. There is worse work to be had. I think of the girls and women claiming livelihoods in Bangladesh's garment industry, and remember again those who died in the collapse of Rana Plaza and the angry labor protests on its anniversary with marching and mourning families who have only received meager compensations from the multinational clothing companies that profit from unregulated labor. I believe that going to a safe school is a human good that girls struggling with real world patriarchies should be able to access. Pushing education is so much better than pushing sterilization. Local NGOs may be able to make good use of the flow of funds triggered by the Girl Effect. I know that elite liberal rhetoric does not mirror on-the-ground realities and politics. I remember the person I met who worked at Intel, who, after hearing me discuss the Girl and seeing the Intel image of a girl holding a Photoshopped sign saying "I am an emerging market," confesses that they have been keeping a secret folder of all the disturbing things that happen inside the sexist world of IT work. Watching the sidewalk of people, myself including, brush past the canvassers, often without even acknowledging their greetings, Toronto crowds begin to look apathetic. For a moment, the options seem to be caring in the wrong way or not caring at all.

If the Girl is a potent phantasmagram that aligns liberal feminist aspiration and sympathy with financial feeling, the turn to correct the generic version of the Girl Effect with precision risk analysis only amplifies the dreamscape that believes capitalism can repair the world with its own logics. I want other hopes for number, for phantasy, and for feminisms. Such hopes are unaligned. They refuse the double forecasts of the invested productive life or the unproductive apocalyptic future. How is it possible to feel the future in other ways?

In the phantasmagram the Girl conjures, capitalism captures desires for a better world. It pulls hopeful affect in, correcting and aligning the world with the imaginable futures that do not challenge capitalism itself, turned into pretty maps with data visualization techniques. Thus, one way to critically evaluate the Girl is to draw attention to the capacities of capitalism to make charged and forceful phantasmagoria, to capture imaginaries, and to suck dreams inside itself. One might be inclined to see the affects that animate the Girl as emanating from the practices and logics that give it quantitative and anticipatory shape. Or, one might be tempted to see capital as driving all wants, as conjuring the very possibility of wanting. Instead, paying attention to phantasmagrams asks us to turn our glance sideways and notice that the phantasy is not so chained to the quantification, that instead imaginaries and hopes were already there, circulating, in excess, doing things with the numbers, as much as numbers were stimulating feelings. There is no absolute functionalist alignment between dreaming futures and financial logics. Capacities to dream and feel other futures are a collective condition that exceed the practices that make up quantification and the surround of capitalism. While calculative infrastructures participate in dreaming the future, there are still many wells of unaligned dreaming that capital fails to register, and that, therefore, make other future imaginaries possible.

Population is used as a neutral term that abstractly describes a multitude: a group of individuals and the total number of inhabitants. In the late twentieth century, when scientists were looking for language to replace *race* as a kind of biological grouping, *population* offered a term seemingly devoid of claims to racial difference.[1] Yet figures of massified life, in the forms of multitudes, crowds, and overpopulation, have been persistently racializing figures. Race is the grammar and ghost of population.[2] *Population* offered an epistemological framing of life that was profoundly objectifying and dehumanizing. It facilitated a distanced and managerial gaze toward optimizing the life and death of brown and black bodies as rates over time in need of adjustment. The entwined histories of colonialism, governmentality, and capitalism are very much persistent in population as a problem space, manifest in the bodies and places that have had to bear the problem of population.

Populate as a verb carries with it another ancient meaning, now designated as obscure by the *Oxford English Dictionary*, and traced back to the latin *populo*: to devastate and to lay waste. In this archaic meaning, to populate is to destroy and conquer, and the contemporary uses of *population* as a noun carry this attachment to violence. To materialize people as the managerial noun of population is to expose them to designations of being living forms of waste available for destruction. *Population*, as a contemporary term describing masses, regularly names the totality of people in prisons, or boxes together people living in precarity as a distanced object. It is profoundly entangled with designations of surplus life, of life unworthy, of life contained, of life open to destruction.

The problem of human population in the late twentieth century was tied primarily to questions of *quantity* within nation-states, and hence the crisis of the overpopulation of some and the underpopulation of others. In the early twentieth century, this was posed as the problem of differential fertility, of poor people having more children than the rich, of blacks having more children than whites, of others outnumbering the hegemonic elite. In the mid-twentieth century, the problem of population became one of surplus people in "developing countries" holding back the economic possibility and life chances of future others. Population was a bomb, in which too many of the wrong kinds of life threatened to destroy economies and the world. In the late twentieth century, population as a problem of economy was financialized, sorting between productive and unproductive life, between life worthy of investment that accumulates value and life not worth being born that, if lived, would diminish the value of the whole.[3] As produced through the demographic transition model, population has been used to legitimate many kinds of dehumanizing calculations of differential life worth.

Today the demographic transition model is still used to make policy, even though subsequent empirical scholarship in demography has deeply troubled it. The model, as promoted in Cold War/postcolonial circuits, plotted population along simplistic axes and a universal timeline. It pulled out the variable of human numbers in a closed nation-state container, separating people from the particularity of histories, cultures, and biographies. Moreover, the demographic transition model presumed the national macroeconomy as its natural and universal environment. Empirically, the model does not hold up for the non-European world, and it does not even work for Europe.[4] More recent demographic research has overturned the idea that any given human population has one singular slope of fertility increase or decline; it is instead composed of a multiplicity of different fertility rates, and each of these rates has to be understood as an artifact of measurement, an outcome of slicing people along one axis or another. In academic research, the universal stages of modernization that the demographic transition assumed have been replaced by detailed studies of the histories and cultures of particular places.[5] Demographers even question whether only those variables amenable to quantification and counting are relevant to demographic processes; other nonquantifiable processes like the work of race, gender, colonial regimes, and religion matter too.[6] Moreover, decades of critical scholarship have shown that transnational distributions of poverty and precarity are not a stage of development but a human accomplishment of complex histories. As a policy object, the Cold War demographic transition model erased the ways class structures the world, instead lumping all people in a nation-state together. As a policy object it is a massive abstraction created through a logic that rendered the world as if it worked like a closed economic system. Its simplicity—materialized in the urgency of the steep upward slope of a line graph—contributed to its success as a potent phantasmagram of policy.

After my time in the archives of postcolonial thick data produced by the experimental exuberance of family planning in Bangladesh, after reading piles of studies about preventing births in the libraries of evidence that the economization of life has created, population has become for me an intolerable concept. It is hard to be against the term *population*, because the concept is so built into the epistemological structures of policy and rule. Yet it is possible, and I think necessary, to be against population. I want better concepts for naming aggregate life.

As a hegemonic figure of aggregate life, population is knotted into the perpetuation of global infrastructures that unevenly invest in and abandon life.[7] Population points the finger at masses rather than distributions and accumulations, at people rather than economy. With climate change, the problem of overpopulation is recharged for the left as well as liberal politics. Photos of global slums and crowded shopping malls in contemporary news media invite viewers to attach to overpopulation as the problem of a world overinhabited and depleted.[8] The earth-altering accomplishment of climate change has reinstalled the epistemic trigger on the population bomb, and overpopulation once again is becoming a way of fearing a world of too many. Africa and South Asia are particularly plotted as the sites of human excess. The phantasy of simply reducing human numbers is so attractive because it does not require the rearrangement of all the other world orders, and particularly the orders of too much accumulation that have accreted in sites with low fertility rates, such as North America, Europe, and East Asia. In the name of earthly life within the horizon of climate change, one can even be against humanity as a whole, a gesture at once against speciesism (that would put humans above other living beings), and a gesture where life in earthly aggregate exceeds, and even diminishes, the human, offering yet another formulation of some must die so that others might live.

Global human numbers have increased dramatically in the last century. More than this, humans as organized through capitalism and industrialism have materially rearranged the planet. The chemical emissions produced by the material accumulation of the few are altering the atmosphere, the oceans, and the very fabric of life for everyone. What kind of population control practices and racisms are reactivated by pointing the finger at human density in a moment when wealthy human–capital assemblages with often low levels of fertility are responsible for the vast bulk of emissions?[9]

The urgency of planetary environmental change is dragging the problem of population — typically narrated as the problem of too many black and brown people — back into the limelight. Africa does not have a particularly high population density, nor has it been a global center of capitalist accumulation, and yet the problem of population is increasingly spatialized as occurring here, as a future of too much African human life in need of preemptory deflection. Contemporary well-known academics from the left are calling out population as a problem that climate change demands returning to, "an elephant in the room" or a "tainted" topic wrongfully avoided.[10] Yet reanimating the problem of population carries with it a thick historical web of infrastructure and epistemology (that this book describes). This infrastructure, far from being moribund, is intent on designating and managing surplus life for the sake of capital. More than this, the problem of population deflects from the crucial fact that it is the structures of industrial accumulation and consumption, justified by the goal of improving macroeconomic measures, that have overwhelmingly produced the material violence of climate change and intensive planetary pollution. Governing industry for the sake of the macroeconomy is a setup that produces molecular material "waste" emissions as outside of the calculation of value, on the one hand, and the governing of human "waste" as integral to the surround that capitalism demands, on the other hand. These two infrastructures of waste-making have worked together to create the decades of purposeful polluting, exposure to violence, and diminishing conditions for the unvalued many.

Cherry picking "population" out of the archive of Cold War logics as the way to address massive ecological change will only serve to revitalize an elite transnational managerial infrastructure that reifies global racisms. As a non-innocent way of materializing a problem, the category of population works to reinstall the supranational world orders of governance — the IMF, World Bank, UN — that have so effectively promulgated neoliberal practices at a planetary scale. In countries with low fertility rates, population policy takes the form of a new era of pro-natalism intent on maintaining the presence of high-consumption, more-valued lives, while simultaneously enforcing borders to limit the arrival of less valued lives from other territories. Converting the problem of climate change and the Anthropocene to the problem of population thereby puts the burden of fixing the world onto rearranging the reproduction of poor and precarious people with highly constrained mobilities. In so doing, the "problem of population" parallels the logic of the Girl Effect, where the burden of saving GDP is placed on the backs of young poor brown girls, once again reinstalling the order of supernational governance as it is. The rich—who often have low fertility, and who profit from planetary environmental damage—are thus let off the hook. I do not have faith that there is a way to remobilize and rework the concept of population so as not to reinvigorate designations of surplus life in this historic moment when racist nationalisms abound. I do not believe that a radical political imaginary for population can be mobilized without amplifying existing infrastructures already being deployed toward more racist necropolitical ends. What politics and concepts are necessary to address climate change as a kind of infrastructural violence of waste-making bound up with the economization of life? We can do better than *population*.

To be against population, however, is not a call for particularizing and individualizing reproductive politics. Liberal feminism in the late twentieth century has successfully installed framings of reproductive choice and rights as an ethicization of family planning in opposition to population control. Feminist reproductive health became a counterpolitic that arranges contraceptive technologies and reproductive services in other, better ways, especially in conditions where access is actively withheld, stigmatized, or even criminalized.[11] Yet, as I hope this book has shown, infrastructures of choice are also central to the history of the economization of life and inventive of the very terms of neoliberal governmentality. Infrastructures built narrowly in the name of individual choice have been repeatedly coupled to the selective minimization of supports to life. Systems of minimized supports provided services or technologies that selectively mitigated conditions, without disrupting them. In the United States and Canada, a deadly racist arrangement of minimized support manifests in high black and indigenous infant mortality rates. Antiblack and anti-indigenous infrastructures distribute killings, overpolicing, incarceration, toxic exposures, and inadequate housing amid the exuberance of individualized choice and commodity spectacle.[12] Yes, the widely dispersed late twentieth-century infrastructures of reproductive health have materially created new affirming possibilities through contraceptive devices and clinical services that people (including myself) can and do use to their own ends. Yet the narrowing of a feminist reproductive politics to choice and consumption repeatedly fails to address the racist and economized infrastructural distributions of value that unevenly diminish and assist the possibilities for sustaining life more generally.

Reproductive commodities themselves contain their own nests of diminishing and assisting relations. Increasingly made in supply chain capitalism, the commodity is made possible by distributions of disposable labor and externalized pollution on the one side, and the reproducible and investable consumer on the other. And yet the act of consumption itself — the taking of the drug, the using of the product, the eating of the food — contains its own rearrangements of risk: the food intervenes in your metabolism and is not merely nutritious, or the product is accompanied by toxicities, or the drug heals as it also has a side effect. The loan is both an investment in life and a rejigging of your precarity. The investment comes in the form of a debt.[13] My consumption is connected to your injury — and even our collective planetary injury. Accumulations and diminishments become distally and unevenly entangled and concentrated. Capitalist biopolitics does not just distribute life and death possibilities between bodies; it bundles antagonistic arrangements of life potential and exposure to death as the very terms of living. The antagonisms, violences, and devaluations are constitutive of the very condition of being alive today.

Over the twentieth century, population and economy, as potent phantasmagrams of aggregate life, became atmospheric as the shared firmament people lived under. They gave form and feeling to capitalism's imaginaries, epistemologies, and infrastructures. Life was surrounded by a world managed for the sake of these two figures. Crucially, however, the infrastructures were full of failures, affects were wily, and epistemologies replete with repressions and contradictions. Subsumption by economy was not a capture of life but rather the constitution of a horizon of expectation that experiments, interventions, and infrastructures would never fully achieve. The economization of life both recomposed and failed to recompose life.

Life remains elusive, in excess of the infrastructures of its management and valuation.

To take a stance against population is not at all a general rejection of contraception, reproductive technologies, experiment, or quantification. It is not a stance against numbers and counting. It is instead to pose the challenge of conjuring other kinds of aggregate forms of life, or better yet *collective forms of life*, that make room for resistance, critique, and the becoming-in-time of multiplicities and relations more responsible to, and less folded within, the violence of capital and its reliance on externalized destructions, racisms, and heterosexual propriety. It *is* possible to experiment and imagine a politics of life that is not biopolitical, that is not invested in the managerial grammar that some must die so that others might live.

Given the necropolitical history of population and the limits of the liberal politics of choice, I suggest that reproduction needs to be retheorized, yet again, to critically account for the ways living-being has, and is, decomposing and recomposing in capitalist formations, as well as to provoke alternative redistributive imaginaries of the one with the many. It was only in the 1980s that *reproduction* became a word tethered and narrowed to questions of individual child birth and fertility in technical worlds. In the eighteenth and nineteenth centuries, it was a word used to describe the becoming-in-time and survival of larger assemblies, such as of species or the relations of production. I think we can rework reproduction to conceptualize how collectivities persist and redistribute into the future and to query *what* gets reproduced. In contrast, population as a figure of aggregate life has been concerned with the governance of quantity and quality, foreclosing questions of the infrastructural *distribution* of life chances, pasts, and futures. Population suppresses the possibility of a politics of redistribution. Better concepts can be found to describe, study, and politicize the problem of human density as it is entangled in the infrastructural and uneven distribution of waste-making and accumulation. The "problem" requires flipping from the question of how much and which *bodies* get to reproduce to what *distributions* of life chances and what kinds of infrastructures get reproduced. *Distributed reproduction* names this better than *population*.

Distributed reproduction is the extensive sense of existing over time that stretches beyond bodies to include the uneven relations and infrastructures that shape what forms of life are supported to persist, thrive, and alter, and what forms of life are destroyed, injured, and constrained. If reproduction is a distributed process of living-being already transformed by racism, birth control, heteronormativities, biomedicine, colonialism, patriarchies, legacies of slavery, pollution, property, development, militarization, financialization, criminalization, nation-states, climate change, industrialized agriculture, labor relations, ecologies, feminism, queer politics, decolonization, love, fear, and hope, then an ontological politics of reproduction is required that can critically grapple with these manifold extensions, with how life is constituted and persists amid these contradictory material relations in ways that recognize, but also exceed, bodies as such. If reproduction is the becoming-in-time of life with the many, distributed reproduction hopes to name the spatially and temporally uneven arrangements of the potentials and relations of this becoming.[14]

In the 1990s, reproductive justice political movements, as an expression of antiracist feminisms, turned the feminist politics of reproduction from a concern of individuals back into the collective concern for a community.[15] Reproductive justice is the struggle for the collective conditions for sustaining life and persisting over time amid life-negating structural forces, and not just the right to have or not have children. Reproductive justice is thus inseparable from environmental justice, antiracism, and anticolonialism. In this spirit, what might a reckoning with distributed reproduction encompass?

Distributed reproduction critically points to the often contradictory and contested relations making up becoming-in-time with the many. It names the tension-filled knots of relations that arrange capacities to persist or alter beyond the flesh of bodies and out into infrastructures, ecologies, epistemologies, and imaginaries. The concept helps me to think both critically and otherwise about not just aggregates but about the more-than-life conditions and histories that compose the world and life chances. Importantly, distributed reproduction is not a romantic conceptualization of flourishing togetherness. Reproduction itself is not a good; rather, it is a process of supporting some things and not others. The conceptualization of distributed reproduction strives to reckon with this fraught process of becoming-in-time that has been constituted through violence, uneven accumulations, and abandonments, and is not merely an affirmation of life. It stretches beyond bodies, individuals, or heterosexuality into the more-than-human, more-than-biotic relations that have been recomposed in the aftermath of capitalism, the nation-state, and macroeconomy. The distributed relations of reproduction can be critically reimagined as already making up a complex non-innocent matrix of life and death accumulations and abandonments to which many are attached, though always unevenly so. Political attention stretches to the historically and spatially extensive infrastructures of technoscience, governmentality, and political economy that do not just converge on but are themselves also the very processes of reproduction.

Learning from the reproductive justice work of organizations like Sister-Song, Asian Communities for Reproductive Justice, Toronto's Native Youth Sexual Health Network, Nayakrishi Andolon, and UBINIG, distributed reproduction might be theorized as a constellation of geopolitically extensive, unevenly distributed, more-than-life "relations of reproduction" that unevenly disperse and rearrange potentials for life in a world riven by capital flows, racialized geographies, environmental destructions, sexual violence, wars, imperialisms, and nation-states—but not only. Such a retheorization participates in provoking a politics that does not concentrate on the body as its only scale (with liberal rights and biomedical access as solutions) and expands beyond the traditional Marxist feminist focus on revaluing reproductive labor. It participates in rejecting population and the horizon of economy, precisely by attending to the conditions and historicity of their operation and reaching for other ways of apprehending what is accumulated, what is maintained, and how. A reformulated sense of distributed reproduction might offer connections and solidarities across the discrepancies of supply chain capitalism, and across the infrastructures that value and devalue life. In other words, I think distributed reproduction has a chance of contributing to the critique of capitalism and the project of imagining and struggling for other worlds at a moment when the material recomposition of life is both local and planetary in scope. Crucially, the politics of distributed reproduction is temporally extensive, stretching to include collectivities that are, that were, and that might be.

Any theorization of distributed reproduction, as much as it may offer a critical diagnostic of life in capitalism, is also a historical symptom. It must wrestle with the ways technical language is caught within the violent histories of persistent forms of doing aggregate life: race, nation, population, and economy, as well as ecology, society, and culture. The relations that a politics of distributed reproduction must grapple with are plural, characterized by collisions and antagonisms, non-innocence and contradictions, in which life can be simultaneously fostered and abandoned, reassembled and destroyed by virtue of inhabiting multiple and conflicting relations. A recharged theorization of reproduction might have a chance at being a crucial political concept because it pivots around becoming—the struggle to exist again but differently, a struggle for the material and conceptual relations of life that is openly connected to the entire earth but also reaches down into the smallest substrates of existence and its interdependencies. It creates an open-ended question in need of continuous critical work. It refuses the simplistic gaze of the managerial elite. Distributed reproduction, as the uneven and yet shared relations of becoming with the many, is bigger than capitalism itself. It surrounds capitalism.

This retheorization of our very surround involves letting go of the phantasy that women, sex, and family might provide special generative possibilities for escape from capitalism, and instead theorizes extensive more-than-life becoming as both already and never fully subsumed. I hope for a politics that might proliferate attention to the temporalities of uneven accumulations, distributions, and possibilities. I hope for a politics that can proliferate increasing exposures to convivial ways of being. I am against population and for concepts that propagate a politics of differently distributed futures.[16]

This book has attempted a history of distribution reproduction within a particular regime of valuation, the economization of life. In this book, distributed reproduction was as much in the exuberant production of experiments, as much in the infrastructures that were built, maintained, and repurposed, as much in recirculation of numbers and calculative practices, as much in the atmospheres of affect as it was in bodies. In this book, distributed reproduction has been manifest as a history of what gets reproduced in the name of population and economy. Historicizing the economization of life in this way has been about what persists, what is destroyed, and what is recomposed. If production names the generativity of the economy, then distributed reproduction, at its most extensive, names the larger variegated process of becoming with the many into the future that stretches beyond "economy" to include the making, breaking, and remaking of life worlds. If so much of life is already materially violated and recomposed in the twenty-first century, then a politics of distributed reproduction must recognize that life is already in the aftermath, which is an ongoing aftermath. More than that, it must actively work to interrupt the continued valuation of precarious life as surplus life, thereby open to more violence.

A politics against population that refuses to reanimate the archive of biopolitical algorithms may well require another apprehension of *life*. Nayakrishi Andolon embraces a conception and ethic of life that at once draws on ecology, experiment, Buddhism, and Sufi Islam. The world is replete with small ways of caring for our distributions.[17] But even the refusal of hegemonic concepts that capitalism does not want to live without is a generative act.[18] What becomes possible if one breaks the bottle of economy and refuses the curve of population? I am not such an optimist as to argue that a radically other, untangled from history, affirmative form of life is possible. I am not such a pessimist that I do not want more than the critical diagnosis. Instead, I hold out for a concept of collective being that can recognize both violence and the possibility of exceeding that violence. That might begin with acts of refusal, including the refusing of concepts, life otherwise, life materialized in other ways, life exceeding our materializations. I can almost imagine a politics, both more than pessimistic and less than optimistic, that refuses, disrupts, diagnoses, redistributes, regenerates, and alters, that is responsible to the uneven and vexed relations of becoming, and does not reproduce the same. I can almost feel it.

Notes

1   Raymond Pearl, "Biology of Population Growth," and *The Biology of Population Growth*. Pearl's curve was indebted to the statistical work of Alfred Lotka (see Sharon Kingsland, *The Evolution of American Ecology*). See also Sabine Höhler, *Spaceship Earth in the Environmental Age*.

2   On the earth as a finite container, see Höhler, *Spaceship Earth*.

3   See Margaret Sanger, *Proceedings of the World Population Conference*. On this conference, see E. Ramsden, "Carving up Population Science." Comments from the participants from Japan, China, Siam, India, Argentine, Chile, Peru, and Brazil were largely absent in the official transcripts of the conference. Exceptions to this silence were a supplementary paper submitted by Rajani Kanta Das, an Indian economist researching at the League of Nations' International Labor Office, and a comment offered in discussion by Kiyo Sue Inui from the Tokyo University of Commerce.

4   On the history of progressivism, feminism, and projects of racial uplift within the spectrum of eugenics, see Childs, *Modernism and Eugenics*; Hustak, "Radical Intimacies"; Kline, *Building a Better Race*; M. Mitchell, *Righteous Propagation*; Paul, *Controlling Human Heredity*; Pernick, *The Black Stork*; Stern, *Eugenic Nation*.

5   On the global history of eugenics, see M. B. Adams, *The Wellborn Science*; Ahluwalia, *Reproductive Restraints*; Campbell, *Race and Empire*; Davie, *Poverty Knowledge in South Africa*; Dubow, *Scientific Racism in Modern South Africa*; Fruhstuck, *Colonizing Sex*; Hodges, "Indian Eugenics in an Age of Reform"; Kevles, *In the Name of Eugenics*; McLaren, *Our Own Master Race*; Park, "Bodies for Empire"; Stepan, "The Hour of Eugenics"; Stern, *Eugenic Nation*.

6   Pearl's relation to eugenics has been debated in the historiography. The portrayal of Pearl as simply anti-eugenics, and thus anti-racist, has been revised to attend to the subtleties of Pearl's critique as existing *within* the scope of eugenics, his continued interest in the scientific management of human breeding, and his racist views (see Allan, "Old Wine in New Bottles"; Barkan, *The Retreat of Scientific Racism*; Mezzano, "The Progressive Origins of Eugenics Critics"). I think, however, the historical account of eugenics in this period needs to be even fur-

ther complicated. The overwhelming focus on the variety of biological racist and antiracist science, while critically necessary, has tended to obscure the rising importance of economic logics coming out of eugenics in the middle of the century. For an alternative account of how biological anti-racism became racialized economic development, see Gil-Riaño, "Historicizing Anti-Racism."

7   As Margaret Schabas (*The Natural Origins of Economics*) has shown, the inauguration of political economy in the eighteenth century by such figures as Carl Linneaus, Adam Smith, and Thomas Malthus initially wove together questions of political economy with those of natural history. It was only later in the nineteenth century that economics and biology became distinct disciplines with divergent practices and domains of inquiry.

8   In fact, Pearl was important to the setting up of the statistical infrastructure at the Department of Agriculture.

9   An example is the theory of stages of social evolution of Lewis Henry Morgan, which influenced the work of Friedrich Engels and Karl Marx (Engels, *The Origin of the Family, Private Property, and the State*). On stages and time in anthropology, see Fabian, *Time and the Other*.

10   On the history of treating Africa as a colonial laboratory, see Tilley, *Africa as a Living Laboratory*.

11   Pearl, *The Biology of Population Growth*, 53, 51.

12   Pearl, *The Biology of Population Growth*, 60.

13   Pearl, *The Biology of Population Growth*, 54.

14   Pearl, *The Biology of Population Growth*, 49.

15   My definition of the term *economization* differs from that offered by Koray Çalışkan and Michel Callon in their well-known essays (Çalışkan and Callon, "Economization, Part 1," and "Economization, Part 2." Çalışkan and Callon use it in a general sense to describe the assembly of actions and devices as "economic," broadly speaking. This stream of sociology of economy draws on actor-network theory to trace how economic phenomena — such as markets — are materialized in technical-human assemblies. Katherine Kenny ("The Biopolitics of Global Health") offers another financialized angle into the economization of life, along the lines of Çalışkan and Callon, in her excellent dissertation work on Disability Adjusted Life Years (DALYs) that resonates with the investment practices described in arc III. In contrast to these uses, this book is defining the term in a narrower way, to specifically historicize practices assigning value for the sake of the macroeconomy. It is thus building on the scholarship of Timothy Mitchell, Susan Bergeron, and others who have historicized the emergence of the macroeconomy in this period (see Bergeron, *Fragments of Development*; Goswami, *Producing India*; Kalpagam, "Colonial Governmentality and the 'Economy'"; T. Mitchell, "Fixing the Economy," and "The Work of Economics"; Suzuki, "The Epistemology of Macroeconomic Reality"). In calling the economization of

life a "regime of value," I am leaning on, and deviating from, Arjun Appadurai's ("Introduction: Commodities and the Politics of Value") discussion of commodities as circulating through multiple regimes of value — in which value here is both economic and cultural. Here, I am concerned with how focusing so intensely on commodities, capital, and labor as the forms through which value is created and contested leaves underinterrogated the work of other historically particular economic formations of value-making. My use here of "regime of value" is also akin to the Foucauldian notion of a "regime of truth" in which the practices and objects by which economy is adjudicated and valued are historically specific. I use the term *regime of value* to mark my interest in historicizing all the components of the economization of life as a mutually constitutive and contingent assemblage.

16 Here I am building on the work of Neferti Tadiar ("Life-Times in Fate Playing"), who describes how valuations of lifetimes and life chances have become a dominant mode within capitalism. Here, this argument, like hers, is interested in examining how value is produced through life beyond the conventional Marxist formulation of the labor theory of value.

17 Drawing on Michel Foucault ("Governmentality"), *governmentality* as a term names practices for directing subjects and objects that develop in states but extend well beyond them. Thus, governmentality is not the equivalent of state governance and instead names the practices that exceed the state and can be found in many other institutions. Foucault's work on biopolitics as a kind of governmentality, and on neoliberal governmentality, are crucial inspirations for this work (Foucault, *History of Sexuality*, Vol. 1, *An Introduction*; *Society Must Be Defended*; *Security, Territory, Population*; *The Birth of Biopolitics*). However, Foucault's own work on neoliberal economics refuses to engage with colonial and postcolonial histories, the elaboration of the racial state, and drops sex as a central analytic. I unfaithfully build on this work in defining economization as a kind of governmentality for the sake of the economy, by which capitalism conjures its own milieu that emerged in racialized and sexualized Cold War and postcolonial itineraries. On theorizing governmentality and its colonial and postcolonial forms, see P. Chatterjee, *The Politics of the Governed*; Kalpagam, "Colonial Governmentality and the 'Economy'"; Scott, "Colonial Governmentality"; Stoler, *Race and the Education of Desire*.

18 In conceptualizing infrastructure here, I am indebted to Anand, "Pressure"; Penny Harvey, *Roads*; Simone, "People as Infrastructure"; Star, "The Ethnography of Infrastructure"; Star and Ruhleder, "Steps towards an Ecology of Infrastructure"; Tousignant, "Insects-as-Infrastructure."

19 For histories of "population" between demographics and ecology, see Bashford, *Global Population*; Höhler, *Spaceship Earth*; Kingsland, *The Evolution of American Ecology*.

20 The potency of "economy" as a phenomenon multiplies the elements found in *material-semiotic actor*, a term widely used in science and technology studies (STS), to include the affective and the infrastructural.

21 In particular, I have built on the rich scholarship of many other scholars who have also taken up the task of historicizing population (Ahluwalia, *Reproductive Restraints*; Bashford, *Global Population*; Connelly, *Fatal Misconception*; Greenhalgh and Winckler, *Governing China's Population*; Hartmann, *Reproductive Rights and Wrongs*; Hodges, *Contraception, Colonialism and Commerce*; Szreter, "The Idea of Demographic Transition"; UBINIG, *Violence of Population Control*) and of grappling with numbers and their enchantments (Day, Lury, and Wakeford, "Number Ecologies"; Desrosières, *The Politics of Large Numbers*; Kalpagam, *Rule by Numbers*; D. M. Nelson, *Who Counts?*).

22 Murphy, *Seizing the Means of Reproduction*.

23 I have used the term *postcolonial* juxtaposed with Cold War in this book to identify the local/transnational interplay of histories of colonization, decolonization, and new imperialisms that characterized so many, but not all, of the places in which U.S.-supported "family planning" programs were created. However, the economization of life, overall, is not necessarily postcolonial, and can just as easily be traced in the history of ongoing settler colonialism in the twentieth century. My emphasis on postcolonial here is largely an artifact of my focus on circuits of expertise between Bangladesh and the United States.

24 The interwar period saw a large literature on optimum population (e.g., East, "Food and Population"; Fairchild, "Optimum Population"), a concept that has recently made a resurgence, such as through the UK organization Optimum Population Trust, and in economics.

25 Pearl, "Differential Fertility."

26 Pearl, *The Biology of Population Growth*, 116.

27 Pearl, *The Biology of Population Growth*, 113.

28 Allan, "Old Wine in New Bottles."

29 McClintock, *Imperial Leather*.

30 My focus here is on the United States, as the largest funder of population control in postcolonial locations. Related histories track the economization of life in communist states, particularly China (Greenhalgh and Winckler, *Governing China's Population*; M. Thompson, "Foucault, Fields of Governability, and the Population").

31 Rose, *The Politics of Life Itself*.

32 Such examples of this feminist work include Briggs, *Reproducing Empire*; Browner and Sargent, *Reproduction, Globalization, and the State*; Casper, *The Making of the Unborn Patient*; Clarke, *Disciplining Reproduction*; A. Y. Davis, *Women, Race and Class*; Davis-Floyd and Dumit, *Cyborg Babies*; Franklin, *Dolly Mixtures*; Franklin and Roberts, *Born and Made*; Ginsburg and Rapp, *Conceiving the New World*

Order; Haraway, *Modest_Witness@Second_Millennium: FemaleMan_Meets_ OncoMouse*; Helmreich, *Alien Ocean*; Jordanova, "Interrogating the Concept of Reproduction in the Eighteenth Century"; Landecker, *Culturing Life*; Mamo, *Queering Reproduction*; J. Morgan, *Laboring Women*; Rapp, *Testing Women, Testing the Fetus*; C. Roberts, *Messengers of Sex*; D. Roberts, *Killing the Black Body*; Rudd, "The United Farm Workers Clinic in Delano, Calif."; Strathern, *Reproducing the Future*; C. Thompson, *Making Parents*; Waldby and Cooper, "The Biopolitics of Reproduction"; Waldby and Mitchell, *Tissue Economies*.

33 Cooper, *Life as Surplus*; Dumit, *Drugs for Life*; Fortun, *Promising Genomics*; Franklin, *Dolly Mixtures*; Franklin, "Ethical Biocapital"; Hayden, *When Nature Goes Public*; Helmreich, "Species of Biocapital"; Landecker, *Culturing Life*; Sunder Rajan, *Biocapital*; Thompson, *Making Parents*; Vora, *Life Support*; Waldby and Cooper, "From Reproductive Work to Regenerative Labour"; Waldby and Mitchell, *Tissue Economies*.

CHAPTER 01 | ECONOMY AS ATMOSPHERE

1 Ann Crittenden, "Why Does a 'Healthy' Economy Feel So Bad?" *New York Times*, July 10, 1977; "A Chilled Economy Feels a 'Breath of Spring,'" *New York Times*, March 29, 1970; Landon Thomas, "Suddenly, a Hesitation about Splurging: A Jittery Economy Stirs Second Thoughts about Ostentation," *New York Times*, September 19, 2007; "Sidelights: Mixed Feelings on Economy," *New York Times*, March 13, 1959; Michael C. Jensen, "Home Buyers All over U.S. Feel the Economy's Crunch," *New York Times*, August 25, 1974, 448.

2 Inspirational for this argument is Brian Massumi's ("National Enterprise Emergency" and *The Power at the End of the Economy*) work on the affective fact.

3 My attention to atmospheres and surrounds is in conversation with Peter Sloterdijk's (*Terror from the Air*) provocative argument that the twentieth century was an era of spheres, in which the dominant European technical and epistemological gesture became the explication of atmosphere, whether this be the literal atmosphere of air as in the examples of gas warfare, the explication of environments as engineerable containers, or the sense of culture as a collective atmosphere. Here, I am arguing the national economy became explicated as a container of relationships open to technical intervention, conjuring a collective macroeconomic atmosphere.

4 On the history of the macroeconomy as an epistemic thing, see M. S. Morgan and Knuuttila, "Models and Modelling in Economics," and Suzuki, "The Epistemology of Macroeconomic Reality." On Keynes's use of statistics, see Werle, "More than a Sum of Its Parts."

5 On the history of economic modeling, see M. S. Morgan, *The World in the Model*; M. S. Morgan and Knuuttila, "Models and Modelling in Economics."

6 The analysis that markets are performativity made through the theories of eco-

nomics and the technologies and practices that make exchange possible has been well developed by sociologists in STS. For example, see Callon, *The Laws of the Markets*; Callon, Millo, and Muniesa, *Market Devices*; McKenzie, Muniesa, and Siu, *Do Economists Make Markets?*

7   Despite the fact that he never went to India, Indian currency was the subject of his first book. Keynes was involved in setting monetary policy for the rupee (Cristiano, "Keynes and India"; Goswami, *Producing India*; Keynes, *Indian Currency and Finance*). Keynes also took a stance relative to famine relief and inoculation, which he considered to be a tempting but economically wrong cosmopolitan humanitarianism. While Keynes held that Malthus's concern for overpopulation did not apply to Britain, or other civilized countries, where prosperity had become hitched to decreases in fertility and where production grew beyond arithmetic increase, he still held that Malthusian population crisis still did apply to India. Since lower population and greater prosperity were tied together for Keynes, death by famine or plague that reduced population would lead to greater wages later, while keeping "skeletons alive" would lead to greater future suffering. Keynes, who held racist and eugenicist positions, believed in a competition between races for land and resources in which he took a decidedly colonialist side in favor of the "English" (Toye, *Keynes on Population*).

8   Keynes, *The General Theory of Employment, Interest and Money*.

9   See "The Economic Possibilities for Our Grandchildren," in Keynes, *Essays in Persuasion*.

10 Toye, *Keynes on Population*, 205.

11 Toye, *Keynes on Population*, 101.

12 For a history of national accounts in the United States and as a global standard, see Carson, "The History of the United States National Income and Product Accounts"; Mitra-Kahn, "Redefining and Measuring the Economy since the Year 1600"; Speich, "The Use of Global Abstractions"; Vanoli, *A History of National Accounting*.

13 Edelstein, "The Size of the U.S. Armed Forces during World War II"; Keynes, *The Economic Consequences of the Peace*; Kuznets, "National Income."

14 Gilbert, "War Expenditures and National Production." On the history of how the GDP is calculated in the U.S, see U.S. Department of Commerce, "GDP."

15 Gilbert and Bangs, "Preliminary Estimates of Gross National Product," 9.

16 On the history of how the GDP is calculated in the United States, see U.S. Department of Commerce, "A Guide to the National Income and Product Accounts of the United States."

17 U.S. Department of Commerce, "GDP."

18 Samuelson and Nordhaus, *Economics*, 402.

19 Landefeld, Seskin, and Fraumeni, "Taking the Pulse of the Economy."

20 Beginning in 2014, GDPNow is produced by the Atlanta Federal Reserve through a technique they call "nowcasting" (Higgins, *GDPNow*).

21 This description of what GDP does and does not count is inspired by Marilyn Waring's (*If Women Counted*) classic feminist analysis.

22 U.S. Department of Commerce, "A Guide to the National Income and Product Accounts of the United States."

23 Fixler and Grimm, "GDP Estimates," 213.

24 Here, I am in dialogue with scholarship on cultures of quantification and number. See Appadurai, *Fear of Small Numbers*; Baucom, *Specters of the Atlantic*; Bouk, *How Our Days Became Numbered*; Day, Lury, and Wakeford, "Number Ecologies"; Guyer, *Marginal Gains*; Igo, *The Averaged American*; D. M. Nelson, *Who Counts?*; McKittrick, "Mathematics Black Life"; Porter, *Trust in Numbers*; Verran, "Number as an Inventive Frontier in Knowing and Working Australia's Water Resources."

25 In thinking through the phantasmagram in the Cold War period, I am drawing inspiration from the work of Joseph Masco (*The Nuclear Borderlands* and "'Survival Is Your Business'"), which demonstrates that one of the crucial accomplishments of Cold War U.S. nuclear and military techniques and infrastructures were powerful new imaginaries and affective orientations. Similarly, Jackie Orr's (*Panic Diaries*) work shows the "psychopolitics" as the affectively distributed panic of Cold War cybernetics and militarism. These works are explicitly dealt with in arc II.

26 I am indebted to the work of Diane Nelson ("Banal, Familiar, and Enrapturing") for thinking number and enchantment together.

27 I will be using the term *phantasy*, rather than *fantasy*, for two reasons. First, I want to use *phantasy* to indicate the extrasubjective origins of the phenomena I am describing. In psychoanalysis, such as the work of Melanie Klein, phantasies come from the unconscious and then invest the external world with significance. In my use of *phantasy* here, the unconscious is not confined to the subject but is generated external to the subject in the extensive epistemic infrastructures, which produce imaginaries that are extraconscious. I want to distinguish phantasy from delusion or irreality, in that phantasy imbues the world with significance and thus is extraobjective, not antagonistic to objects. Second, the term *phantasy* is consistent with the use of the term *phantasma* and the history of phantasmagoria, and thus traces an itinerary through Walter Benjamin's theorizations of the spectral productions of capitalism.

28 Zizek, *The Sublime Object of Ideology*; Boyle, "The Four Fundamental Concepts of Slavoj Žižek's Psychoanalytic Marxism."

29 Birla, *Stages of Capital*.

30 Pateman, *The Sexual Contract*.

31 Samuelson, *Economics*. Interestingly, Simon Kuznets originally wanted to include labor within the home. This patriarchal exclusion was fought for and won.

32 Christophers, "Making Finance Productive."

33 F. Cooper and Packard, "Introduction."

34 Speich, "Traveling with the GDP through Early Development Economics' History"; M. Morgan, "Seeking Parts, Looking for Wholes."

35 Clark, *The Conditions of Economic Progress*.

36 Speich, "The Kenyan Style of 'African Socialism': Developmental Knowledge Claims and the Explanatory Limits of the Cold War."

37 In the last decade there has been a proliferation of the language of "dashboards" to describe the use of a multiplicity of indices.

38 Fukuda-Parr, "Rescuing the Human Development Concept from the HDI: Reflections on a New Agenda."

39 Ul Haq, *Reflections on Human Development*, 5.

40 See United Nations Development Programme, "Multidimensional Poverty Index," accessed July 10, 2016, http://hdr.undp.org/en/statistics/mpi/.

41 See Millennium Project, "Goals, Targets and Indicators," accessed July 10, 2016, http://www.unmillenniumproject.org/goals/gti.htm.

42 Costa and James, *The Power of Women and the Subversion of Community*; Federici, *Wages against Housework*.

43 Coined by Selma James; Costa and James, *The Power of Women and the Subversion of Community*.

44 See James, *Sex, Race and Class*.

45 Jordanova, "Interrogating the Concept of Reproduction in the Eighteenth Century."

46 J. Roger, *Buffon*.

47 For some discussions of the historicity of reproduction and its refiguration with Charles Darwin, see Grosz, "Darwin and the Ontology of Life"; and Mueller-Wille, "Figures of Inheritance, 1650–1850."

48 Marx, *Capital*, Vol. 1, chapter 23, 620.

49 Marx, *Capital*, Vol. 1, chapter 23, 633.

50 Marx, *Capital*, Vol. 1, chapter 15, sec. 3A, 431.

CHAPTER 02 | DEMOGRAPHIC TRANSITIONS

1 Notestein, "Abundant Life."

2 Dillon, *Composite Report of the President's Committee to Study the United States Military Assistance Program*, 71.

3 Draper's committee was influenced by the work of William Vogt, Henry Fairfield Osborn Jr., and Hugh Moore.

4 Draper, quoted in "Making Foreign Aid Work," *New York Times*, July 26, 1959.

5 The language of the population bomb was developed by Hugh Moore in his

pamphlet (*The Population Bomb*, 1959) and amplified later by Paul Ehrlich in his best-selling book (*The Population Bomb*, 1971).

6 Draper joined up with population control enthusiast and businessman Hugh Moore to advocate for a "Manhattan Project" for population, which included racist paid advertising about population and crime, as well as the explicitly anti-Communist 1954 pamphlet *Population Bomb*. For more on their endeavors, see Merchant 2015.

7 See, for example, the history of the Carolina Population Center, the founding of Family Health International (FHI), and the International Fertility Research Program—all examples of USAID-funded nonprofits that could undertake controversial programs at a slight distance from the state. For more on this formation, see arc II and Murphy, *Seizing the Means of Reproduction*.

8 On this history of the demographic transition, see Caldwell, "Demographers' Involvement in Twentieth-Century Population Policy"; Szreter, "The Idea of Demographic Transition and the Study of Fertility."

9 Notestein's model had three stages. However, the middle stage was most commonly split into two further stages, for a total of four. Notestein's model was based on the findings of the Princeton European Fertility Project concerned with the nineteenth century, the conclusions of which have become hotly contested in demography. He was also influenced by modernization theory and the work of Walt Whitman Rostow (*The Stages of Economic Growth*).

10 For a summary of these contestations, see Szreter, "Theories and Heuristics."

11 Engerman et al., *Staging Growth*; Latham, *Modernization as Ideology*; Gilman, *Mandarins of the Future*.

12 Today, the demographic transition has a fifth stage, correlated to population decline in Europe. On the biopolitics of population decline, see M. Cooper, "Resuscitations"; Waldby and Cooper, "The Biopolitics of Reproduction."

13 Notestein, "Population—The Long View."

14 On the rise of postcolonial experts, see McCann, "Malthusian Men and Demographic Transitions"; Mehos and Moon, "The Uses of Portability"; T. Mitchell, *Rule of Experts*.

15 Khan, "Policy-Making in Pakistan's Population Programme," 31.

16 On Mahalanobis's statistical work on caste and intelligence, see Setlur, "Searching for South Asian Intelligence."

17 V. K. R. V. Rao, *The National Income of British India*.

18 For descriptions of the Pakistani bureaucracy and civil service, see Burki, "Twenty Years of the Civil Service of Pakistan"; Hull, *Government of Paper*; Papanek, *Pakistan's Development*.

19 On Pakistan bureaucracy and Islamabad, see Hull, *Government of Paper*.

20 Burki, "Twenty Years of the Civil Service of Pakistan"; F. Ahmed, "Pakistan Forum."

21 Papanek, *Pakistan's Development*.

22 For texts from the HAG that described Pakistan as a field laboratory, see Papanek, *Pakistan's Development*; Papanek and Mujahid, "Effect of Policies on Agricultural Development." On the notion of "natural experiments" in economics, see T. Mitchell, "The Work of Economics."

23 Curle, *Planning for Education in Pakistan*, 1.

24 Hence, the title of T. Mitchell, *Rule of Experts*. See also Mehos and Moon, "The Uses of Portability."

25 On colonial and postcolonial sites as living laboratories, see Tilley, *Africa as a Living Laboratory*. One might also compare this history with that of Chile, another laboratory of U.S. economists, the "Chicago Boys" (Arditti, Klein, and Minden, *Test-Tube Women*; Fornazzari, *Speculative Fictions*; Silva, "Technocrats and Politics in Chile").

26 Notestein, "Abundant Life," 321.

27 Foucault, *Society Must Be Defended*.

28 Coale and Hoover, *Population Growth and Economic Development in Low-Income Countries*.

29 Coale and Hoover, *Population Growth and Economic Development in Low-Income Countries*, 24.

30 Watkins, "If All We Knew about Women Was What We Read in *Demography*, What Would We Know?"; Mackinnon, "Were Women Present at the Demographic Transition?"; Janssens, "'Were Women Present at the Demographic Transition?'"

31 On the history and politics of population control, see Ahluwalia, *Reproductive Restraints*; Briggs, *Reproducing Empire*; Connelly, *Fatal Misconception*; Crichtlow, *Intended Consequences*; Greenhalgh and Winckler, *Governing China's Population*; Hartmann, *Reproductive Rights and Wrongs*; Hartmann and Standing, *Food, Saris and Sterilization*; Kligman, *The Politics of Duplicity*; Mamdani, *The Myth of Population Control*; M. Rao, *From Population Control to Reproductive Health*; Robertson, *The Malthusian Moment*; Silliman and Bhattacharjee, *Policing the National Body*.

32 For example, see N. Chatterjee and Riley, "Planning an Indian Modernity."

33 On this history, see Bashford, *Global Population*; Merchant, "Prediction and Control."

34 Meadows et al., *The Limits to Growth*, 115.

CHAPTER 03 | AVERTED BIRTH

1 Enke, "The Economic Aspects of Slowing Population Growth," and "Birth Control for Economic Development."

2 Connelly, *Fatal Misconception*; Johnson, "Address in San Francisco at the 20th Anniversary Commemorative Session of the United Nations."

3   Enke, "The Economic Aspects of Slowing Population Growth," 47.

4   Enke, "The Economic Case for Birth Control," 41.

5   Population Council, "Taiwan: Births Averted by the IUD Program." On the history of the Taiwan IUD program, see Takeshita, *The Global Biopolitics of the IUD*.

6   Guttmacher Institute, "2.3 Million Births Averted in Korea, Savings Huge."

7   For example, see Barrett, "Potential Fertility and Averted Births"; Keyfitz, "How Birth Control Affects Births"; Malitz, "The Costs and Benefits of Title XX and Title XIX Family Planning Services in Texas"; Mauldin, "Births Averted by Family Planning Programs"; McCarthy, "Contraceptive Sterilization in Four Latin American Countries"; Nortman, "Status of National Family Planning Programmes of Developing Countries in Relation to Demographic Targets"; Potter, "Additional Births Averted When Abortion Is Added to Contraception"; Potter, "Births Averted by Contraception"; Venkatacsharya and Das, "An Application of a Monte Carlo Model to Estimate Births Averted Due to Various Family Planning Methods."

8   Mauldin, "Births Averted by Family Planning Programs."

9   Potter, "Additional Births Averted When Abortion Is Added to Contraception."

10   Saunders, "Research and Evaluation."

11   Saunders, "Research and Evaluation."

12   Malitz, "The Costs and Benefits of Title XX and Title XIX Family Planning Services in Texas"; Kramer, "Legal Abortion among New York City Residents."

13   Enke, "An Economist Looks at Air Force Logistics."

14   TEMPO II provided a more complex and long-range simulation of population and economic dynamics featuring such variables as government spending and education. TEMPO II was meant to be a tool planners would actually use to model their development budgets and planning choices. Yet significantly USAID preferred TEMPO I, as the simpler model emphasized the importance of reducing population, which was "a principle objective" of USAID's, while TEMPO II diluted this variable in a more complex model (Walsh, "Evaluation of the GE-TEMPO Project").

15   Moreland et al., "The Carolina Population Center Family Planning Administrator Training Game."

16   Moreland et al., "The Carolina Population Center Family Planning Administrator Training Game."

17   National Academy of Sciences, *Rapid Population Growth*, 4.

18   National Security Council, *Implications of Worldwide Population Growth for U.S. Security and Overseas Interests*, 8.

19   Benjamin, *The Arcades Project*; Buck-Morss, *The Dialectics of Seeing*.

20   McCann, "Malthusian Men and Demographic Transitions."

CHAPTER 04 | DREAMING TECHNOSCIENCE

1 On Begum Roquiah, see Hasanat, "Sultana's Utopian Awakening"; Hossain, "The Begum's Dream"; Ray, *Women of India*; Sarkar, *Visible Histories, Disappearing Women*.

2 Hossein, "Sultana's Dream," 31.

3 Hossein, "Sultana's Dream," 38, 39.

4 Hossein, "Sultana's Dream," 37.

5 Hossein, "Sultana's Dream," 37.

6 Hossein, "Sultana's Dream," 33.

7 Hossein, "Sultana's Dream," 46.

8 Cohen, "Ethical Publicity"; Mukharji, "Swapnaushadhi."

CHAPTER 05 | INFRASTRUCTURES OF COUNTING AND AFFECT

1 National Institute of Population and Training (NIPORT), "Bangladesh Demographic and Health Survey," 185.

2 NIPORT, "Bangladesh Demographic and Health Survey," 197.

3 NIPORT, "Bangladesh Demographic and Health Survey," 11.

4 Chaaban and Cunningham, *Measuring the Economic Gain of Investing in Girls*, 16.

5 "Snapshot."

6 NIPORT, "Bangladesh Demographic and Health Survey," 11.

7 I use the term *thick data* to describe the dense enumeration practices in the twentieth century. I want to purposely contrast thick data with big data. Postcolonial thick data is entangled with the history of twenty-first-century big data, from projects to regulate the flow of labor to biosecurity, and hence new forms of empire. This point is made in arc III. We could also, however, extend this analysis to the thick data of settler colonialism, and look at the way counting and audit is used as a kind of fiscal colonialism against Indigenous people. See Pasternak, "To 'Make Life' in Indian Country."

8 Girl Effect, "Assets and Insights."

9 NIPORT, "Bangladesh Demographic and Health Survey," 22.

10 NIPORT, "Bangladesh Demographic and Health Survey," 13, 12, 161.

11 On the politics of counting as encounter, see the excellent work of Day, Lury, and Wakeford, "Number Ecologies"; D. M. Nelson, "Banal, Familiar, and Enrapturing"; Verran, "Number as an Inventive Frontier in Knowing and Working Australia's Water Resources."

12 I am indebted to Alexandra Widmer's ("Of Temporal Politics and Demographic Anxieties," and "Unsettling Time") understanding of the relation between conscripts of modernity, as theorized by David Scott (*Conscripts of Modernity*), and demographic practices.

13 Masco, *The Theater of Operations*. On Cold War rationality, see Erickson et al.,

How Reason Almost Lost Its Mind; Gilman, *Mandarins of the Future*; Latham, *The Right Kind of Revolution*.

14   Orr, *Panic Diaries*.

15   Fanon, *The Wretched of the Earth*, "Medicine and Colonialism," and "The North African Syndrome."

16   Fanon, *Black Skin, White Masks*, 3–16.

17   R. L., "Wanderings of the Slave"; Sexton, "The Social Life of Social Death."

18   On infernal alternatives, see Stengers, Pignarre, and Goffey, *Capitalist Sorcery*.

19   My argument here has important parallels to Li's (*The Will to Improve*) work on the ways neoliberal governmentalities of development operate through enticements.

20   Population Council, *A Manual for Surveys of Fertility and Family Planning*.

21   Mani, *Contentious Traditions*; Sinha, *Specters of Mother India*.

22   On the history of the survey, see Igo, *The Averaged American*.

23   Stycos, "Survey Research and Population Control in Latin America," 368; also see Warwick, "The KAP Survey."

24   Simon, "A Huge Marketing Research Task," 24. Julian Simon would later make a radical switch from enthusiastically bringing free market and marketing approaches to the population problem to famously making an about-face and taking a position against population control programs, arguing that most demographic modeling was not specific enough and did not take into account the benefits of technology, innovation, and infrastructure that larger populations can induce, as well as arguing against claims of a depleting planet as presented by Paul Ehrlich (*The Population Bomb*) and Meadows et al. (*Limits to Growth*).

25   Warwick, "The KAP Survey."

26   Population Council, *A Manual for Surveys of Fertility and Family Planning*.

27   Wyon and Gordon, *The Khanna Study*, xviii. On the history of the Khanna study, see Connelly, *Fatal Misconception*; Williams, "Rockefeller Foundation Support to the Khanna Study."

28   Ravenholt, "World Fertility Survey," 2.

29   National Security Council, *Implications of Worldwide Population Growth for U.S. Security and Overseas Interests*.

30   The count of twenty-eight is based on Population Council (1970) figures from *A Manual for Surveys of Fertility and Family Planning* (Population Council), and includes unpublished studies.

31   The count of four hundred comes from a remark by Everett Rogers (*Communication Strategies for Family Planning*, xvii) about the overwhelming number of studies in India.

32   Population Council, *A Manual for Surveys of Fertility and Family Planning*, 189–220.

33   I am here drawing on the work of many scholars who have already written about

the history of family planning in South Asia (Ahluwalia, *Reproductive Restraints*; Amrith, *Decolonizing International Health*; Arnold, *Colonizing the Body*; N. Chatterjee and Riley, "Planning an Indian Modernity"; Chattopadhayay-Dutt, *Loops and Roots*; Fannin, "Domesticating Birth in the Hospital"; Hodges, *Reproductive Health in India*, and *Contraception, Colonialism and Commerce*; Mamdani, *The Myth of Population Control*).

34 Bandarage, *Women, Population and Global Crisis*; Connelly, *Fatal Misconception*; Ehrlich, *The Population Bomb*; Höhler, *Spaceship Earth*.

35 National Security Council, *Implications of Worldwide Population Growth for U.S. Security and Overseas Interests*, 75.

36 National Security Council, *Implications of Worldwide Population Growth for U.S. Security and Overseas Interests*, 7–8.

37 National Security Council, *Implications of Worldwide Population Growth for U.S. Security and Overseas Interests*, 61.

38 Population Council, *A Manual for Surveys of Fertility and Family Planning*; Stycos, "Survey Research and Population Control in Latin America"; Warwick, "The KAP Survey."

39 Examples include Khanna, Senegal, Comilla, Sialkot.

40 Stix and Notestein, *Controlled Fertility*.

41 Coale and Hoover, *Population Growth and Economic Development in Low-Income Countries*. See also Walle, "Fertility Transition, Conscious Choice, and Numeracy."

42 Thus, contraception was sold through what Lauren Berlant (*Cruel Optimism*) calls an optimistic attachment.

43 I discuss USAID's supply-side approach and its antagonistic relations to feminist practices in my *Seizing the Means of Reproduction*.

44 Ravenholt and Gillespie, "Maximizing Availability of Contraception through Household Utilization."

45 Ravenholt, "The Power of Availability."

46 S. Ahmed, "Affective Economies."

47 Patel, "Advertisements, Proprietary Heterosexuality, and Hundis."

48 Mohan, *Advertising Management*.

49 For one of the PSI founder's accounts, see P. D. Harvey, *Let Every Child Be Wanted*. The Social Marketing Project (SMP) of Bangladesh became the Social Marketing Company and split from PSI, which is based in Washington, DC (Bhandari, "Communications for Social Marketing"; Davies, Mitra, and Schellstede, "Oral Contraception in Bangladesh"; P. D. Harvey, "Advertising Affordable Contraceptives"; Luthra, "Contraceptive Social Marketing in the Third World"; Schellstede and Ciszewski, "Social Marketing of Contraceptives in Bangladesh").

50 This point resonates with the Deleuzian argument made by Maurizio Lazzarato, who explains that by buying labor force, capital does not buy the subject of the worker for a given time but instead the right to "exploit a 'complex' assemblage" that includes "transportations, urban models, media, and ways of perceiving and feeling, as well as semiotic systems" (*Signs and Machines*, 43). Relatedly, I am arguing that enrolling people into family planning acceptance was a project of installing and then exploiting a complex assembly of experiment, consumption, counting, markets, and feeling as a condition for attaching to the promises of the economy that would make value from surplus life.

51 The KAP surveys sit in juxtaposition to "affective labor," in which creativity, thought, passion, fear, affection, affinity, intimacy, aspiration, sensation, and so on became prominent generative forces valued as a kind of labor in the contemporary. Scholars who study affective labor trace paid forms (ranging from the care work of the nanny to the passionate sociality of entrepreneurs) as well as unpaid forms (from parenting to forming affectively charged communities on social media) that produce economic opportunity in the form of new markets, ideas, and even forms of life. As such, affectively charged activities can be harvested to stimulate economic value even if they are not directly waged, such as the way online social networks generate data that can be used to target advertising. The proliferation of the KAP surveys points to another itinerary for affect and its entanglements with capitalism in the late twentieth century.

52 For example, the framework of Coale and Hoover's landmark study, *Population Growth and Economic Development*, universalized Khanna findings and then modeled their application to Mexico.

53 Mamdani, *The Myth of Population Control*.

54 Mamdani, *The Myth of Population Control*, 19.

55 Mamdani, *The Myth of Population Control*, 23.

56 In a chapter draft written in 1864 and not originally included in *Capital*, Vol. 1, Marx distinguishes between formal and real subsumption. In formal subsumption, arrangements of labor are brought into capitalist relations without substantially changing. Workers become compelled to work for wages for their subsistence. In real subsumption, capitalism works to rearrange its own conditions; capital transforms forms of life into capitalist forms that become thoroughly imbued with the nature and requirements of capital. In real subsumption, we might say that capital works to continually take apart and rearrange itself in all the nooks and crannies of life in order to extract value from the world. Michael Hardt and Antonio Negri, in *Empire*, develop this line of thinking by theorizing the "real subsumption of society" as the absorption of all life and society into capital (see Makdisi, Casarino, and Karl, *Marxism beyond Marxism*).

1   On motivation in family planning, see the forthcoming doctoral works of Savina Balasubramanian on family planning motivation in India and Kira Lussier on the history of motivation theory in labor, as well as Parry, *Broadcasting Birth Control.*

2   Demeny, "Observations on Population Policy and Population Program in Bangladesh," 314.

3   Rogers, *Communication Strategies for Family Planning.*

4   Demeny, "Observations on Population Policy and Population Program in Bangladesh," 311.

5   Rogers, *Communication Strategies for Family Planning.*

6   Coughlin, "Male Sterilization"; Krishnakumar, "Ernakulam's Third Vasectomy Campaign Using the Camp Approach."

7   Rogers, *Communication Strategies for Family Planning.*

8   Dhanraj, *Something like a War.*

9   Ahmad, "Field Structures in Family Planning"; Khan, Chowdury, and Bhuiya, "An Inventory of the Development Programmes by Government and Non-Government Organizations in Selected Unions of Matlab (Excluding BRAC–ICDDR,B)"; Robinson, "Family Planning in Pakistan 1955–1977."

10 Khan, Chowdury, and Bhuiya, "An Inventory of the Development Programmes by Government and Non-Government Organizations in Selected Unions of Matlab (Excluding BRAC–ICDDR,B)."

11 Robinson, Shah, and Shah, "The Family Planning Program in Pakistan."

12 Ahmad, "Field Structures in Family Planning"; Cleland and Mauldin, "The Promotion of Family Planning by Financial Payments"; Enke, "The Gains to India from Population Control," and "The Economic Aspects of Slowing Population Growth"; Kangas, "Integrated Incentives for Fertility Control"; Rogers, "Incentives in the Diffusion of Family Planning Innovations."

13 Enke, "Reducing Fertility to Accelerate Development," and "The Economic Aspects of Slowing Population Growth."

14 USAID, "Project Paper: Population/Family Planning, Bangladesh," 63.

15 Incentives were also critiqued almost as soon as they were suggested. For example, Notestein was against them, worrying that "it is quite possible that to poor and harassed people financial inducements will amount to coercion and not to an enlargement of their freedom" ("Abundant Life," 828).

16 Kangas, "Integrated Incentives for Fertility Control," 1283.

17 Berelson, "Beyond Family Planning," 8.

18 Donors' Community in Dhaka, "Position Paper on the Population Control and Family Planning Programme in Bangladesh"; see also Connelly, *Fatal Misconception*; Hartmann, *Reproductive Rights and Wrongs.*

19 Ministry of Health and Population Control, *Population Control Programme in Bangladesh.*

20 Akhter, *Depopulating Bangladesh*; Hartmann, *Reproductive Rights and Wrongs*; Hartmann and Standing, *Food, Saris and Sterilization*; UBINIG, *Faces of Coercion*, and *Violence of Population Control.*

21 The minimization of supports was also done as a strategy of increasing contraceptive spread. For example, family planning services (including pharmaceutical versions) were moved out of clinics; no prescription, no interaction with medical personnel (instead all sorts of agents might disperse contraception), and no follow-up were necessary. Age expanded to include young and old. These tactics might be compared with the ways big pharmaceutical companies have expanded the market for antidepressants (Dumit, *Drugs for Life*).

CHAPTER 07 | EXPERIMENTAL EXUBERANCE

1 Appadurai, "Number in the Colonial Imagination," and *Fear of Small Numbers*; Cohn, *Colonialism and Its Forms of Knowledge*; Goswami, *Producing India*; Gupta, *Red Tape*; Hull, *Government of Paper*; Kalpagam, *Rule by Numbers*; Nazar Soomro and Murtaza Khoso, "Impact of Colonial Rule on Civil Services of Pakistan"; Prakash, *Another Reason*; Raman, *Document Raj*; Setlur, "Searching for South Asian Intelligence."

2 Population Control and Family Planning Division, *Innovative Projects in Family Planning and Rural Institutions in Bangladesh*, 3.

3 Population Control and Family Planning Division, *Innovative Projects in Family Planning and Rural Institutions in Bangladesh*, 4.

4 Population Control and Family Planning Division, *Innovative Projects in Family Planning and Rural Institutions in Bangladesh*, 4.

5 These examples are from the charity Save the Children and the Mothers' Clubs of Bangladesh's Rural Social Service Project of the Department of Social Welfare (Population Control and Family Planning Division, *Innovative Projects in Family Planning and Rural Institutions in Bangladesh*, 22–23, 36–37).

6 On off-the-grid infrastructures, see Anand, "Pressure"; Simone, "People as Infrastructure"; Street, *Biomedicine in an Unstable Place.*

7 Arnold, *Science, Technology and Medicine in Colonial India*, and *Colonizing the Body*; Bonneuil, "Development as Experiment"; Tilley, *Africa as a Living Laboratory.*

8 Nguyen, "Government-by-Exception," and *The Republic of Therapy.*

9 On experimentality in drug trials, see M. Cooper and Waldby, *Clinical Labor*; Dumit, *Drugs for Life*; Kelly and Geissler, *The Value of Transnational Medical Research*; Peterson, *Speculative Markets*; Sunder Rajan, *Biocapital*. On experimentality in medical research, see Geissler, *Para-States and Medical Science*. On

experimentality and family planning in South Asia, see Cohen, "Operability, Bioavailability, and Exception"; Towghi, "Normalizing Off-Label Experiments and the Pharmaceuticalization of Homebirths in Pakistan"; Towghi and Vora, "Bodies, Markets, and the Experimental in South Asia."

10 Murphy, *Seizing the Means of Reproduction* 2012.

11 H.-J. Rheinberger, *Toward a History of Epistemic Things*, 5. See also Rheinberger, "Experimental Systems," and "Infra-Experimentality." On experiment and reproduction in South Asia, see Bharadwaj, "Experimental Subjectification"; Towghi, "Normalizing Off-Label Experiments and the Pharmaceuticalization of Homebirths in Pakistan"; Towghi and Vora, "Bodies, Markets, and the Experimental in South Asia"; Vora, "Indian Transnational Surrogacy and the Commodification of Vital Energy."

12 This argument is moving the definition of *experiment* out of the laboratory setting, away from truth claims, and toward a kind of engagement with the world. I think it is important to understand experiment, which has its roots in the word *experience*, as a more extensive practice toward inducing something different to happen, and thus experimental forms like clinical trials, or laboratory experiments that want replicability, are just a subset of a larger range of practices that qualify *experimental*. This wider definition acknowledges the fact that experiments can be just as much about arranging the conditions as about creating knowledge. So many so-called experiments — such as big pharma clinical trials or corporate research setups toward measuring the toxicity of something — are not actually set up to produce surprising findings or new rearrangements of the world but instead are oriented to preserving the conditions under which corporates operate. Thus this proposed definition of *experiments* also suggests we might push back against the scientific integrity of corporate research and validate projects that aspire to change conditions already shaped by hegemonic infrastructures. However, this definition places *experiments* in the contradiction of simultaneously eliciting the changeability that capitalism desires for itself and the alteration toward an otherwise that might be antagonistic to its founding conditions.

13 Ali, *Planning the Family in Egypt*; Bledsoe, *Contingent Lives*; Cohen, "Ethical Publicity," and "Operability, Bioavailability, and Exception"; Kaler, *Running after Pills*; Pande, *Wombs in Labor*; Rapp, *Testing Women, Testing the Fetus*; Tarlo, *Unsettling Memories: Narratives of the Emergency in Delhi*; Van Hollen, *Birth in the Age of AIDS*, and *Birth on the Threshold*.

14 Lorde, *A Burst of Light*, 131.

15 Self-help projects and cooperative movements were a significant feature of re-imagining the terms of "rural development" in the 1970s. Some of these, as other scholars have documented, also ended up being remobilized as social enterprise projects within the neoliberal inventiveness of the moment. Both the Grameen

Bank and BRAC have such roots. See Hasnath, "The Practice and Effect of Development Planning in Bangladesh."

16 Here, we might differentiate between work that collects interviews in order to re-present them as critiques, such as the work of UBINIG, and research that is part of larger assemblies that seeks to add feminist consideration of women's voices or gender to public health and family planning research.

17 Asad, "Conscripts of Western Civilization?," 337; see also D. Scott, "Colonial Governmentality."

18 On the impossibility and necessity of reckoning with genocide through numbers, see D. M. Nelson, *Reckoning*.

19 "Disaster; East Pakistan: Cyclone May Be the Worst Catastrophe of Century," *New York Times*, November 22, 1970; Reilly, "Great Bhola Cyclone, 1970."

20 On the war, see Mascarenhas, *Bangladesh*; Mookherjee, "The 'Dead and Their Double Duties'"; Raghavan, *1971*; Saikia, *Women, War, and the Making of Bangladesh*, 19.

21 On war rape, see D'Costa, *Nationbuilding, Gender, and War Crimes in South Asia*, and "Women, War, and the Making of Bangladesh"; D'Costa and Hossain, "Redress for Sexual Violence before the International Crimes Tribunal in Bangladesh"; K. S. Islam, "Breaking Down the *Birangona*"; Menen, "The Rapes of Bangladesh," *New York Times*, July 23, 1972; Mookherjee, "The Absent Piece of Skin," "Gendered Embodiments," "'Remembering to Forget,'" and *The Spectral Wound*; Saikia, *Women, War, and the Making of Bangladesh*.

22 The number of dead in the war is certainly a high and horrific figure, though it is also a highly debated and contentious issue. Whatever the calculation, the war, and the longer period between 1970 and 1975, was a period of extreme mass death in Bangladesh. The Hamoodur Rahman Commission Report (October 23, 1974), an official Pakistan government investigation, put the war death figure as low as twenty-six thousand civilian casualties. However, this low number is largely rejected by scholars. The history of the three million figure accepted by the government of Bangladesh is less clear. On his blog, journalist David Bergman, Dhaka-based editor of the paper *New Age*, has discussed the possible origins of this number in a chain of misquoting in newspaper reporting in 1971, which was then popularized by Sheikh Mujib in a well-known speech on January 10, 1972. For his posts from 2011, the International Crimes Tribunal of Bangladesh (established by the government which began in 2009) has charged Bergman with contempt. In 2015 forty-nine people signed a public statement titled "Bergman er Shajai Ponchash Nagoriker Udbek" (Statement of concern regarding tribunal's contempt judgment on David Bergman), published in the *Daily Prothom Alo* on December 20, 2014. Farida Akhter, a internationally respected feminist intellectual and activist whose work is discussed in this book, was among the signatories. Bina

D'Costa, a well-known scholar who has published widely on war rape in Bangladesh, has also been cited for contempt for her work on the trial. These contempt rulings underline how discussions of the war and its war tribunal are politically contentious.

Some significant accounts of the war dead are: (1) the 2008 estimate by Obermeyer et al. in *British Medical Journal* is 269,000; (2) an estimate by Curlin, Chen, and Hussain (1976) based on Matlab data gives a national estimate of 500,000; (3) M. A. Hasan, the convener of the War Crimes Facts Finding Committee, has created estimates based on excavations of mass graves, calculating there are 500,000 buried in mass graves, which represents about 30 percent of the dead, leading to a figure of 1.8 million; (4) Samilla Bose in her controversial and widely criticized book estimates the war dead at 50,000—100,000. Bose, *Dead Reckoning*; Curlin, Chen, and Hussain, "Demographic Crisis"; Hasan, "Discovery of Numerous Mass Graves, Various Types of Torture on Women"; Obermeyer et al., "Fifty Years of Violent War Deaths from Vietnam to Bosnia."

23 On mass-grave counts, see Hasan, "Discovery of Numerous Mass Graves, Various Types of Torture on Women."

24 Contempt charges have been issued by the International Crimes Tribunal, a hotly contested tribunal, with both criticism and support caught in the antagonism between the ruling and opposition parties. Since it is a domestic tribunal, it can only charge Bangladeshis with crimes. Among those charged and executed for war crimes have been members of the Jamaat-e-Islami party, the largest Islamist party in Bangladesh. The trials have been accompanied by accusations of government interference, protests, hacked Skype conversations of the judge, and international censure of its practices.

25 The final refugee total as of December 15, 1971, determined by the Indian government. See Vogler, "The Birth of Bangladesh," 28.

26 UNHCR, "The State of the World's Refugees 2000," 60.

27 Alamgir, *Famine in South Asia*; Razzaque, "Sociodemographic Differentials in Mortality during the 1974–75 Famine in a Rural Area of Bangladesh"; Razzaque et al., "Sustained Effects of the 1974–5 Famine on Infant and Child Mortality in a Rural Area of Bangladesh."

28 Nurul Islam, *Making a Nation: Bangladesh*; Sobhan, "Politics of Food and Famine in Bangladesh."

29 Sen, *Poverty and Famines*.

30 Mbembe, "Aesthetics of Superfluity," and "Necropolitics."

31 D'Costa, *Nationbuilding, Gender, and War Crimes in South Asia*, and "Women, War, and the Making of Bangladesh: Remembering 1971"; Menen, "The Rapes of Bangladesh"; Mookherjee, "The Absent Piece of Skin," "Available Motherhood," "Gendered Embodiments," "'Remembering to Forget,'" and *The Spectral Wound*; Saikia, *Women, War, and the Making of Bangladesh*; Robert Trumbull,

"Dacca Raising the Status of Women While Aiding Rape Victims," *New York Times*, May 12, 1972.

32   K. S. Islam, "Breaking Down the *Birangona*"; Mookherjee, *The Spectral Wound*; Saikia, *Women, War, and the Making of Bangladesh*.

33   Mookherjee, "Available Motherhood," and "Gendered Embodiments."

34   Planning Commission, *The First Five Year Plan, 1973–78*, 538.

35   Planning Commission, *The First Five Year Plan, 1973–78*, 539.

36   National Security Council, "Implications of Worldwide Population Growth for U.S. Security and Overseas Interests."

37   USAID, "Disaster Relief: Bangladesh, Civil Strife, January–September, 1972," 1.

38   Demeny, "Observations on Population Policy and Population Program in Bangladesh," 308.

39   Das, *Whispers to Voices*, 1.

40   Cons and Paprocki, "Contested Credit Landscapes"; Karim, *Microfinance and Its Discontents*; Yunus and Jolis, *Banker to the Poor*. On gender and microcredit, see Rankin, "A Critical Geography of Poverty Finance."

41   Smilie, *Freedom from Want*.

42   Yunus, "The Problem of Poverty in Bangladesh."

43   According to the PSI website, accessed February 26, 2015, http://www.psi.org/work-impact/countries/.

44   Alvarez, "Advocating Feminism"; Bernal and Grewal, *Theorizing NGOs*; Haider, "Genesis and Growth of the NGOs"; Karim, "Demystifying Micro-Credit"; Lang, "The NGOization of Feminism"; Silliman, "Expanding Civil Society."

45   Nazneen and Sultan, "Struggling for Survival and Autonomy: Impact of NGOization on Women's Organizations in Bangladesh."

46   Fanon, *The Wretched of the Earth*; Scott, "Colonial Governmentality," discusses this splintering as a feature of colonial governmentality.

47   P. Chatterjee, *The Politics of the Governed*; Cohen, "Ethical Publicity."

48   Dean, *Governmentality*; Foucault, *The Birth of Biopolitics*; Rose, "Governing 'Advanced' Liberal Democracies."

49   Grimes et al., "Sterilization-Attributable Deaths in Bangladesh."

50   Cohen, "Operability, Bioavailability, and Exception"; Towghi and Vora, "Bodies, Markets, and the Experimental in South Asia."

51   Clarke, *Disciplining Reproduction*; M. Cooper and Waldby, *Clinical Labor*; Franklin, *Biological Relatives*, and *Dolly Mixtures*; Hartmann, *Reproductive Rights and Wrongs*; Landecker, *Culturing Life*; L. Morgan and Roberts, "Reproductive Governance: Rights and Reproduction in Latin America"; C. Roberts, *Messengers of Sex*; D. Roberts, *Killing the Black Body*; Rose, *The Politics of Life Itself*; C. Thompson, *Making Parents*; Waldby and Mitchell, *Tissue Economies*.

52   Franklin, *Biological Relatives*; Ginsburg and Rapp, *Conceiving the New World Order*; Mamo, *Queering Reproduction*; Taussig, Rapp, and Heath, "Flexible Eu-

genics"; C. Thompson, *Making Parents*; Vora, "Indian Transnational Surrogacy and the Commodification of Vital Energy."

53 Dumit, *Drugs for Life*; Helmreich, "Blue-Green Capital, Biotechnological Circulation and an Oceanic Imaginary," and "Species of Biocapital"; Kloppenburg, *First the Seed*; Peterson, *Speculative Markets*; Sunder Rajan, *Biocapital*; Waldby, "Stem Cells, Tissue Cultures and the Production of Biovalue"; Waldby and Cooper, "The Biopolitics of Reproduction."

54 As scholars such as Melinda Cooper, Catherine Waldby, Kaushik Sunder Rajan, Adriana Petryna, Joseph Dumit, Vinh-Kim Nguyen, Cori Hayden, Kristin Peterson, and others have shown. The international division of labor in transnationalized pharmaceutical research seeks out dispossessed communities from which to draw experimental subjects who are hired for the biological capacities and incapacities of their bodies in response to biomedical interventions and as such bear bodily risks. In the United States, we can trace an economy of clinical research done on poor, racialized, and imprisoned bodies (Abadie, *The Professional Guinea Pig*; Briggs, *Reproducing Empire*; Fisher, *Medical Research for Hire*; Hornblum, *Acres of Skin*; Markowitz and Rosner, *Lead Wars*; Reverby, *Examining Tuskegee*; D. Roberts, *Killing the Black Body*; Washington, *Medical Apartheid*). Over the twentieth century, risk has legally become a property carried by the persons (as determined by trials over insurance claims concerning slaves who had freed themselves on ships), and hence individuals are understood to contract to take on more or less risk as part of their labor and choice (Levy, *Freaks of Fortune*). Opening oneself to biomedical alteration and bearing risk in experiments has become a precarious form of labor, which could be outsourced and arranged in transnational assemblies that could take advantage of unequal regulatory environments and geopolitical distributions of dispossession. Catherine Waldby and Melinda Cooper ("The Biopolitics of Reproduction") argue that this articulation of reproductive life, capital, and technoscience has composed new forms of biolabor. They importantly point out that this labor is often unwaged as part of an ethical orientation that manifests in organ and blood donation policies, patients opting for experimental treatments, or voluntary surrogacy. Hence, experimental arrangements of biocapital are highly stratified by monetized and nonmonetized activity, by uneven distributions of regulation, by racialized and gendered geopolitical arrangements of outsourcing risk, and by hierarchies of speciesism.

55 Lawrence Cohen's ("Operability, Bioavailability, and Exception") work on sterilization and kidney transplant operation in India has importantly traced that to be "operable," on the one hand, is to be assimilated into the norms of modern citizenship through the act of submitting one's body to surgery. To be "bioavailable," on the other hand, is to be that body whose living-being is available for

extraction and redistribution, under the rhetoric of the "charitable gift" that deploys life and hope for the future.

56 Examples of this legitimation of experiments with new contraceptives as a kind of assistance or gift were prevalent in the work of IFRP in the 1970s. See Bhuiyan and Begum, "Quinacrine Non-Surgical Female Sterilization in Bangladesh"; Kessel, "100,000 Quinacrine Sterilizations"; Potts and Paxman, "Depo-Provera—Ethical Issues in Its Testing and Distribution"; Scully, "Maternal Mortality, Population Control, and the War in Women's Wombs."

57 Akhter, *Depopulating Bangladesh*, "Reproductive Rights," and *Resisting Norplant*; Arditti, Klein, and Minden, *Test-Tube Women*; Ehrenreich and Ehrenreich, *The American Health Empire*; Hartmann and Standing, *Food, Saris and Sterilization*; UBINIG, *Faces of Coercion*; *Violence of Population Control*; and *Women and Children of Bangladesh as Experimental Animals*.

58 FINRRAGE and UBINIG, "Declaration of Comilla," vii.

59 Murphy, *Seizing the Means of Reproduction*; B. Rahman, "Expenditures and Funding of Population Programs in Bangladesh." For a history of FHI's presence in Bangladesh, see its website, accessed April 9, 2015, http://www.fhi360.org /countries/Bangladesh.

60 Beyond just family planning research, in the second half of the twentieth century there was also an exuberant investment into materializing and harnessing biological reproduction in the form of micrological living capacities, thereby rendering life a domain open to unprecedented alteration and intervention. Yet the capacities of cells, viruses, DNA, and gametes—recombinatory, animating capacities to generate difference in time—could only perform as forms of biocapital or biotechnology when "assisted" by technoscientific assemblages (Helmreich, "Blue-Green Capital"). "Assisted reproduction," then, was not confined to efforts to biomedically enhance and modify fertility but was found in far more extensive assemblages that "assisted" capacities to both live and die, to be responsive or not, to metabolize and recombine, to change and become otherwise, to persist or decompose. The exuberant experimentality of family planning was just one expression of a dense history of technoscientific efforts to "assist" in generating different forms and arrangements of life. As such, we might extend the sense of "assisted" reproduction to the governmentalities of family planning experiment; to the infrastructures of health care, water, and legal protections that allow some forms of life, and not others, to persist; or even more widely to ecological, agricultural, military, industrial, and state infrastructures that unevenly produce violent and injuring environments that patch together the postindustrial world. Experiment, as an assemblage that conjures futures and recomposes life, is a distributed reproduction.

Beyond assisted reproduction, then, the extensive relations that constitute the capacity of life to become-in-time with others makes up what I call "distributed reproduction" (Murphy, "Distributed Reproduction"). I use this term to recognize that reproduction is not a process confined to individual bodies. Distributed reproduction recognizes the extensive reach of reproduction as a capacity to persist and change beyond flesh into infrastructures and other relations, as a process that is organized in assemblies of bodies, organisms, and ecologies along with their nonorganic supports. Distributed reproduction helps me to think otherwise about aggregate forms of life, and thus the concept is developed in this book's coda.

61 Franklin and Lock, *Remaking Life and Death.*

62 C. Thompson, *Making Parents.*

63 "PM: 100 Economic Zones to Be Set Up," *Dhaka Tribune,* October 21, 2015. http://www.dhakatribune.com/bangladesh/2015/oct/21/pm-100-economic-zones-be-set.

CHAPTER 08 | DYING, NOT DYING, NOT BEING BORN

1 The other rural field sites are in Abhoynagar, Mirsarai, Chakaria, Sylhet, and Mirzapur.

2 In the 1950s and 1960s, the U.S. government had huge amounts of agricultural surpluses, particularly wheat, that it used to establish the Food for Peace program. The program did not gift the food but required receiving countries to pay into a fund in local currency. The money was called PL480 money. Large sums of money accumulated in accounts in Pakistan in the 1960s, as a courted Cold War ally, that then could be used by U.S. programs. This money was used to fund the Cholera Research Laboratory (CRL). CRL's W. Henry Mosley remembers, "In effect it meant that the US government, with the consent of Pakistan, ordered that this money could be spent locally for development activities. Because we were sending over such vast amounts of food grain, huge amounts of this money had accumulated in the country. Certainly from the U.S. side, this was almost considered free money. Anyway, when the Cholera Lab had started going in East Pakistan, so there were some foreign exchange requirements like to pay the salaries of the U.S. scientists and buy some U.S. equipment; there was a dollar appropriation for that, but all the local operations were under the PL480 money. It was literally like an open checkbook. There was no limit to the amount of money you could spend on projects that could be done locally. You could do lots of things. You could hire local people to do these daily rectal swabs and pay salaries. As long as commodities and things could be purchased locally you could have all you wanted . . . some of that equipment was sort of hand-me-downs. Some lab equipment came from NIH that they were upgrading or they weren't using any more. Some very, very good equipment, I remember. I don't know exactly what

that cost in dollar amounts during the East Pakistan day, but probably not very much.... After the War the PL480 moneys from Lahore must have been shut off. When I was there [as director in the 1970s], we were operating entirely under US dollars" (D. Sack, "The Pak-SEATO Cholera Research Laboratory," 6–7).

3   Two volumes looking at their accomplishments and history have been produced by ICDDR,B (Fauveau, *Matlab*; J. Sack and Rahim, *Smriti*).

4   Findings from these successive vaccine trials were published in Benenson, Joseph, and Oseasohn, "Cholera Vaccine Field Trials in East Pakistan"; Clemens, "Impact of B Subunity Killed Whole-Cell and Killed Whole-Cell Only Oral Vaccines against Cholera upon Treated Diarrhoeal Illness and Mortality in an Area Endemic for Cholera"; Curlin, *Immunological Aspects of a Cholera Toxoid Field Trial in Bangladesh*; Mosley et al., "Report of the 1966–67 Cholera Vaccine Field Trial in Rural East Pakistan"; Mosley, Alauddin Chowdhury, and Aziz, "Demographic Characteristics of a Population Laboratory in Rural East Pakistan."

5   Aziz and Mosley, "The History, Methodology and Main Findings of the Matlab Project in Bangladesh"; Mosley, Alauddin Chowdhury, and Aziz, "Demographic Characteristics of a Population Laboratory in Rural East Pakistan."

6   Aziz and Mosley, "The History, Methodology and Main Findings of the Matlab Project in Bangladesh"; van Ginneken et al., *Health and Demographic Surveillance in Matlab.*

7   Mahalanabis et al., "Oral Fluid Therapy of Cholera among Bangladesh Refugees. 1973." Multiple institutions take credit for the invention of ORT. Credit is given to Dr. Dilip Mahalanabis, the physician who led the team working in the refugee camps in India; Johns Hopkins, which ran a Medical Research and Training Center in Calcutta; and Richard Cash and David Nalin of Harvard, who worked with the Cholera Research Laboratory, Harvard, and Johns Hopkins. In turn, BRAC developed the distribution system for ORT with ICDDR,B (Bhattacharya, "History of Development of Oral Rehydration Therapy"; Cash, "A History of the Development of Oral Rehydration Therapy (ORT)"; Ruxin, "Magic Bullet"; Schultz, "From a Pump Handle to Oral Rehydration Therapy").

8   Chowdhury, Vaughan, and Abed, "Use and Safety of Home-Made Oral Rehydration Solutions"; Nalin, "Oral Maintenance Therapy for Cholera in Adults"; A. S. M. M. Rahman et al., "Mothers Can Prepare and Use Rice-Salt Oral Rehydration Solution in Rural Bangladesh."

9   Sixty-five dollars per averted birth is the price Lawrence Summers used in his calculations of the effects of contraception versus education on fertility reduction (Summers, "Investing in All the People," 7). Five cents is the approximate price of an oral rehydration packet (Taylor and Greenough, "Control of Diarrheal Diseases," 228).

10 Data from UNICEF, Under-5 mortality statistics. See UNICEF, "Bangladesh at

a Glance" (2008), accessed April 4, 2015, http://www.unicef.org/bangladesh
/cbg_(18.10.08).pdf, and UNICEF, "Statistics," accessed April 4, 2015. http://
www.unicef.org/infobycountry/bangladesh_bangladesh_statistics.html.

11 On the biopolitics of the minimal fix in humanitarian emergency aid, see Redfield, "Bioexpectations."

12 Tan-Torres Edejer et al., "Achieving the Millennium Development Goals for
Health"; Varley, Tarvid, and Chao, "A Reassessment of the Cost-Effectiveness of
Water and Sanitation Interventions in Programmes for Controlling Childhood
Diarrhoea."

13 Fauveau, "Does ORT Reduce Diarrhoeal Mortality?"

14 Pierce, "How Much Has ORT Reduced Child Mortality?"

15 Guerrant, "Cholera, Diarrhea, and Oral Rehydration Therapy."

16 Fassin, "Humanitarianism as a Politics of Life," and *Humanitarian Reason*; Redfield, *Life in Crisis*; Ticktin, *Casualties of Care*.

17 This language is sprinkled throughout the archive of Matlab studies. For example, see Mosley, Alauddin Chowdhury, and Aziz, "Demographic Characteristics of a Population Laboratory in Rural East Pakistan."

18 Osteria et al., *Assessment of the Matlab Contraception Distribution Project: Implications for Program Strategy*; Phillips et al., "The Demographic Impact of the Family
Planning–Health Services Project in Matlab, Bangladesh."

19 A list of BRAC–ICDDR,B projects is available at "Working Papers of the BRAC–
ICDDR,B Joint Research Project at Matlab," accessed April, 21, 2015, https://www.
icddrb.org/what-we-do/publications/doc_download/3754-summary. See also
Khan, Chowdury, Bhuiya, "An Inventory of the Development Programmes by
Government and Non-Government Organizations in Selected Unions of Mat-
lab."

20 Gilmore, *Golden Gulag*, 28.

CHAPTER 09 | EXPERIMENTAL OTHERWISE

1 Akhter, *Depopulating Bangladesh*; *Resisting Norplant: Women's Struggle in Bangladesh against Coercion and Violence*; "Reproductive Rights"; and *False-Linkage
of Food and Population*; UBINIG, *Faces of Coercion: Sterilization Tearing Apart
Organs*; *Violence of Population Control*; *Women and Children of Bangladesh as Experimental Animals*.

2 Akhter, *Resisting Norplant*.

3 Akhter, *Seeds of Movements*; Hossein, *Sultana's Dream*.

4 Akhter, *Seeds of Movements*, 15.

5 Mazhar, "Nayakrishi Andolon"; Narigranth Prabartna, *On Shahaj Way to Ananda*.

6 UBINIG, "About UBINIG," 2.

7 UBINIG, "About UBINIG," 5.

8 Akhter, *Seeds of Movements*, 285.

9   Mazhar et al., "Uncultivated Food."

10 Farida Akhter, interview with author, January 16, 2013.

CHAPTER 10 | INVEST IN A GIRL

1   The math and figures here are those used in Summer's original speech ("The Most Influential Investment"), even though they do not quite add up.

2   See Summers, "The Most Influential Investment," 5.

3   My thinking on anticipation is indebted to a collaboration with Vincanne Adams and Adele Clarke (V. Adams, Murphy, and Clarke, "Anticipation: Techno-science, Life, Affect, Temporality").

4   Becker, "Investment in Human Capital."

5   Theodore W. Schultz, "The Economics of Being Poor: Noble Prize Lecture." Stockholm, December 8, 1979.

6   Patel, "Risky Subjects: Insurance, Sexuality, and Capital."

7   Barro and Becker, "Fertility Choice in a Model of Economic Growth"; Becker and Tomes, "Child Endowments and the Quantity and Quality of Children."

8   National Security Council, *Implications of Worldwide Population Growth*, 11, 69.

9   For excellent ethnographies of Girl Effect projects, see the work of Hayhurst, "Corporatising Sport, Gender and Development," and Moeller, "Proving 'The Girl Effect.'" See also Koffman and Gill, "'The Revolution Will Be Led by a 12-Year-Old Girl'"; Switzer, "(Post)Feminist Development Fables." In September 2015, the Girl Effect separated from the Nike Foundation and became an independent NGO.

10 Intel, 10x10 advertisement, Mediaplanet insert in *USA*, September 17, 2010, 5.

11 On the fetus as such a figure, see Lauren Berlant, "America, 'Fat,' the Fetus." Thinking with Berlant, in the Girl Effect, the fetus still remains an important site of national and capitalist rejuvenation, but through averting the possible future fetus by installing particular kinds of value in the Girl.

12 Plan, "Because I Am a Girl, Invest in Me." 2009 video, http://www.youtube.com/watch?v=N4ok_5D27BY.

13 Plan Canada, "Because I Am a Girl," April 1, 2010, accessed February 24, 2010, https://www.youtube.com/watch?v=rMErh8luzSE.

14 Plan Canada, "Because I Am a Girl."

15 Thank you to Joe Dumit for sharing this point.

16 "World Economic Forum Annual Meeting 2009," *World Economic Forum* website, January 11, 2011, accessed September 10, 2013, http://www.weforum.org/events/world-economic-forum-annual-meeting-2009.

17 Levine et al., *Girls Count*, xiii.

18 Lawson, "Women Hold Up Half the Sky," 9–10.

19 Smiley, "Interview of Secretary of State Hillary Rodham Clinton," January 27, 2010, accessed September 24, 2016, http://www.state.gov/secretary/20092013clinton

/rm/2010/01/136489.htm. During Clinton's tenure as secretary of state this quote was featured on the U.S. Department of State webpage concerning the International Fund for Women and Girls. This webpage is no longer available. U.S. Department of State, accessed April 20, 2015, http://www.state.gov/s/gwi /programs/womensfund/why/. This webpage is archived here, accessed September 24, 2016, http://web.archive.org/web/20130815014214/http://www.state .gov/s/gwi/programs/womensfund/why/.

20 Laura Bush, radio address, quoted in *Washington Post*, November 17, 2001.

21 Nicholas Kristof, "Smart Girls vs. Bombs."

CHAPTER 11 | EXHAUSTING DATA

1 Cowen, *The Deadly Life of Logistics*, 103.

2 Maplecroft, "Indices," accessed December 10, 2015, http://maplecroft.com /portfolio/ethical_insight/portfolio/mapping/maplecroft/landing/.

3 Maplecroft, "Indices."

4 Alyson Warhurst, Maplecrofts's CEO, developed this approach in her work advising the mining industry, including De Beers. See her speech "Corporate Social Responsibility and the Mining industry: Presentation to Euromines," June 4, 1999, Brussels, accessed December 1, 2013, http://www.dlist.org/sites/default /files/doclib/Corporate%20Social%20Responsibility%20and%20the%20Min ing%20Industry.pdf. See also her 2012 PopTech presentation, "Risk Mapper," accessed December 1, 2013, http://poptech.org/people/alyson_warhurst.

5 Maplecroft, "Advisory Services," accessed December 12, 2015, https://maple croft.com/about/news/womens_girls_right_index.html.

6 UNICEF, Children's Rights and Business Atlas, accessed December 12, 2015, http://www.childrensrightsatlas.org/.

7 See, for example, this language and strategy in the scenario planning Maplecroft did for Nike (Nike Inc., "Sustainable Business Performance Summary").

8 Maplecroft, "Girls Discovered," accessed July 15, 2014, http://girlsdiscovered .org/. In June 2016, Girls Discovered was retooled Girl Stats, and in 2014 Maplecroft was sold to become Verisk Maplecroft.

9 Girl Effect, "Assets and Insights." Data analyzed by Maplecroft. See also Girls Discovered, "Global Data Gaps." On the emphasis on data collection in Girl Effect projects, see Moeller, "Proving 'The Girl Effect.'"

10 Girl Effect, "The State of Girl Data," and "Global Availability of Data on Girls."

11 Girl Effect, "The State of Girl Data."

12 Girl Effect, "The State of Girl Data."

13 Girl Effect, "Data and Development."

14 Girl Effect, "Data and Development," and "The State of Girl Data."

15 Bill Gates, video introduction to annual letter, 2013, accessed December 3, 2013, http://annualletter.gatesfoundation.org/.

16 Gates, "Bridging the Gender Gap," *Foreign Policy*, July 13, 2013.

17 UN Global Pulse Project, http://www.unglobalpulse.org/. See UN Global Pulse White Paper, *Big Data for Development: Opportunities and Challenges* (July 2012), http://www.unglobalpulse.org/.

18 See UN Global Pulse, accessed December 2, 2013, http://www.unglobalpulse .org/, and, "Big Data for Development: Opportunities and Challenges" (UN Global Pulse paper, May 2012), accessed December 2, 2013. http://www .unglobalpulse.org/sites/default/files/BigDataforDevelopment-UNGlobal PulseJune2012.pdf.

19 Girl Effect, "Data and Development," and "Empowering Girls with Economic Assets."

20 See Girl Effect, "Empowering Girls with Economic Assets"; Quisumbing and Kovarik, *Investments in Adolescent Girls' Physical and Financial Assets*. On poverty finance, see Rankin, "A Critical Geography of Poverty Finance."

21 Sebsted, "Girls and Their Money," 11.

22 Lindi Hlanze, economic advisor, UK Department for International Develop- ment (DFID), quoted in Girl Effect, "Empowering Girls with Economic Con- trol."

23 On the Basic Income Grant in South Africa, see Ferguson, *Global Shadows*.

24 For example, Zimmerman, Tosh, and Holmes, "Investing in Girls."

25 Anke Schwittay calls this the "financial inclusion assemblage." Schwittay, "The Financial Inclusion Assemblage."

26 Chaia et al., "Half the World Is Unbanked," Financial Access Initiative Report, US, 2010, accessed December 7, 2013, http://mckinseyonsociety.com/half-the -world-is-unbanked/.

27 Keraan, "Banking the Bottom of the Pyramid."

28 Roy, *Poverty Capital*, and "Subjects of Risk." See also Rankin's discussion of so- cial capital ("Social Capital, Microfinance, and the Politics of Development").

29 Bill and Melinda Gates, "Our Big Bet for the Future," Gates annual letter 2015, accessed June 6, 2016, http://www.gatesnotes.com/2015-Annual-Letter?page=5.

30 Maria A. May, "Mobile Money Needs the Support of Grassroots Organisations to Reach Its Potential," BRAC blog, January 25, 2015, accessed April 17, 2015, http://blog.brac.net/2015/01/mobile-money-needs-the-support-of-grassroots -organisations-to-reach-its-potential/#sthash.pzpEM5KL.dpuf.

31 Anushka Zafar, "Maya Apa: Bringing Information to Every Woman in Bangla- desh," BRAC blog, February 12, 2015, accessed April 17, 2015, http://blog.brac .net/2015/02/maya-apa-bringing-information-to-every-woman-in-bangladesh/; "Snapshot."

32 Still from Intel's 10x10 campaign video "Not a Number," August 24, 2012, ac- cessed April 18, 2013, https://vimeo.com/48163647. Images of girls flash by while captions state, "I am not a number, but I count," followed by stills of girls hold-

ing Photoshopped placards with "I am" statements: "I am a girl," I am capable,"
"I am a great investment," and "I am the emerging market."

33 Here, I am drawing on the arguments about lifetimes and fate playing developed
by Tadiar, "Life-Times in Fate Playing."

CODA

1   For example, one can track the rise of the use of *population* in population genetics
and in epidemiology.

2   This formulation of grammar and ghost comes from Jared Sexton, "The Social
Life of Social Death."

3   See Tadiar, "Life-Times in Fate Playing."

4   For one of the best critical discussions of the demographic transition model
within demography, see Szreter, "Theories and Heuristics." See also Boserup,
Population and Technological Change; Hartmann, *Reproductive Rights and Wrongs*;
Kreager, "Population and the Making of the Human Sciences"; Szreter, Sholkamy, and Dharmalingam, *Categories and Contexts*.

5   See, for example, the work rethinking the demographic analysis in African sites
(Bledsoe, *Contingent Lives*; Cordell and Gregory, *African Population and Capitalism*; Ittmann, Cordell, and Maddox, *The Demographics of Empire*).

6   See Szreter, "Theories and Heuristics."

7   However, one could make a different argument here. Since population is so embedded in infrastructures, one could argue that a powerful strategy is to change
the terms of "population" as a concept to make it do different political and material work. I think that this recuperative strategy will fail because it will feed
already powerful racisms at work.

8   Global Development Professionals Network, "Overpopulation, Overconsumption—In Pictures." *Guardian*, April 1, 2015. https://www.theguardian.com/global
-development-professionals-network/gallery/2015/apr/01/over-population
-over-consumption-in-pictures.

9   Gore, "Extreme Carbon Inequality."

10 See, for example, Haraway, "Anthropocene, Capitalocene, Plantationocene,
Chthulucene"; Chakrabarty, "Climate and Capital."

11 See Murphy, *Seizing the Means of Reproduction*; J. Nelson, *Women of Color and the
Reproductive Rights Movement*; Davis, *The Making of "Our Bodies, Ourselves"*; Morgen, *Into Our Own Hands*; Solinger, *Beggars and Choosers*.

12 See Pasternak, "To 'Make Life' in Indian Country"; Million, *Therapeutic Nations*;
Agathangelou, "Neoliberal Geopolitical Order and Value"; Gilmore, *Golden
Gulag*.

13 We can see this in the DALY (Disability Adjusted Life Years), a global health
measure of loss of health in the standardized quantum of life-time, which facili-

tates the further calculation of cost efficiencies of preserving and not preserving "life-time" itself (Kenny, "The Biopolitics of Global Health").

14 Here I draw on anthropologist Shellee Colen's influential notion of "stratified reproduction," which names how "kinship" is hierarchically rearranged and made geopolitically mobile by structures of race, sex, and class in transnational political economies of circulating labor ("'Like a Mother to Them'"). At a conceptual register, anthropologist Marilyn Strathern (*Reproducing the Future*) has developed a notion of "dispersed kinship" in order to ask complex questions about the shifting range of "procreators" who take part in, "assist," and hence are in "relation to" reproductive acts as mediated by technoscience, property forms, and knowledge production. Feminist technoscience studies scholar Donna Haraway (*When Species Meet*) offers the concept of "mess mates" as part of a project to unwind the Western-bounded human by attending to the ways humans are always something more by virtue of being knotted to the nonhuman others through which living is possible. The concept of distributed reproduction has been developed in Murphy, "Distributed Reproduction," "Distributed Reproduction, Chemical Violence, and Latency," and "Reproduction."

15 See Asian Communities for Reproductive Justice, "A New Vision for Advancing Our Movement for Reproductive Health, Reproductive Rights, and Reproductive Justice"; SisterSong Women of Color Reproductive Health Collective, *Reproductive Justice Briefing Book*; Silliman et al., *Undivided Rights*.

16 On "becoming with the many," see Haraway, "When Species Meet."

17 For thinking about the island of decolonial otherwise that is already here, I draw on Bellacasa, "Making Time for Soil"; L. Simpson, *Islands of Decolonial Love*; Tallbear, "Why Interspecies Thinking Needs Indigenous Standpoints."

18 On refusal and interruption, I am drawing on Chua, "Logistics, Capitalist Circulation, Chokepoints"; Cowen, *The Deadly Life of Logistics*; Pasternak, "The Economics of Insurgency"; A. Simpson, *Mohawk Interruptus*; Tuck and Yang, "R-Words."

Bibliography

Abadie, Roberto. *The Professional Guinea Pig: Big Pharma and the Risky World of Human Subjects*. Durham, NC: Duke University Press, 2010.

Adams, Mark B., ed. *The Wellborn Science: Eugenics in Germany, France, Brazil, and Russia*. New York: Oxford University Press, 1990.

Adams, Vincanne, Michelle Murphy, and Adele Clarke. "Anticipation: Technoscience, Life, Affect, Temporality." *Subjectivities* 28, no. 1 (2009): 246–65.

Agathangelou, Anna M. "Neoliberal Geopolitical Order and Value." *International Feminist Journal of Politics* 15, no. 4 (2013): 453–76.

Ahluwalia, Sanjam. *Reproductive Restraints: Birth Control in India, 1877–1947*. Urbana: University of Illinois Press, 2008.

Ahmad, Wajihuddin. "Field Structures in Family Planning." *Studies in Family Planning* 2, no. 1 (1971): 6–13.

Ahmed, Feroz. "Pakistan Forum: Building Dependency in Pakistan." *MERIP Reports* 29 (June 1974): 17–20.

Ahmed, Sara. "Affective Economies." *Social Text* 22, no. 2 (2004): 118–39.

Akhter, Farida. *Depopulating Bangladesh: Essays on the Politics of Fertility*. Dhaka: Narigrantha Prabartana, 1992.

———. *False-Linkage of Food and Population: The Man-Made Scare for Corporate Solutions*. Dhaka: Narigrantha Prabartana, 2005.

———. "Reproductive Rights: A Critique from the Realities of Bangladeshi Women." *Re/productions*, no. 1, 1998.

———. *Resisting Norplant: Women's Struggle in Bangladesh against Coercion and Violence*. Dhaka: Narigrantha Prabartana, 1995.

———. *Seeds of Movements: On Women's Issues in Bangladesh*. Dhaka: Narigrantha Prabartana, 2007.

Alamgir, Mohiuddin. *Famine in South Asia: Political Economy of Mass Starvation*. Cambridge, MA: Oelgeschlager, Gunn and Hain, 1980.

Ali, Kamran Asdar. *Planning the Family in Egypt: New Bodies, New Selves*. Austin: University of Texas Press, 2002.

Allan, Garland E. "Old Wine in New Bottles: From Eugenics to Population Control in the Work of Raymond Pearl." In *The Expansion of American Biology*, edited by

Keith R. Benson, Jane Maienschein, and Ronald Rainger, 231–61. New Brunswick, NJ: Rutgers University Press, 1991.

Alvarez, Sonia E. "Advocating Feminism: The Latin American Feminist NGO 'Boom.'" *International Feminist Journal of Politics* 1, no. 2 (1999): 181–209.

Amrith, Sunil S. *Decolonizing International Health: Indian and Southeast Asia, 1930–65.* Basingstoke: Palgrave Macmillan, 2006.

Anand, Nikhil. "Pressure: The PoliTechnics of Water Supply in Mumbai." *Cultural Anthropology* 26, no. 4 (2011): 542–64.

Anderson, Dillon, et al. *Composite Report of the President's Committee to Study the United States Military Assistance Program.* 2 vols. Washington, DC: President's Committee, 1959.

Appadurai, Arjun. *Fear of Small Numbers: An Essay on the Geography of Anger.* Durham, NC: Duke University Press, 2006.

———. "Introduction: Commodities and the Politics of Value." In *The Social Life of Things: Commodities in Cultural Perspective,* edited by Arjun Appadurai, 3–63. Cambridge: Cambridge University Press, 1988.

———. "Number in the Colonial Imagination." In *Orientalism and the Postcolonial Predicament,* edited by Carol A. Breckenridge and Peter van der Veer, 314–40. Philadelphia: University of Pennsylvania Press, 1993.

Arditti, Rita, Renate Duelli Klein, and Shelley Minden, eds. *Test-Tube Women: What Future for Motherhood?* London: Pandora Press, 1984.

Arnold, David. *Colonizing the Body: State Medicine and Epidemic Disease in Nineteenth-Century India.* Berkeley: University of California Press, 1993.

———. *Science, Technology and Medicine in Colonial India.* New York: Cambridge University Press, 2000.

Asad, Talal. "Conscripts of Western Civilization?" In *Dialectical Anthropology: Essays in Honor of Stanley Diamond,* vol. 1, edited by Christine Ward Gailey, 333–51. Gainesville: University Press of Florida, 1992.

Asian Communities for Reproductive Justice. "A New Vision for Advancing Our Movement for Reproductive Health, Reproductive Rights, and Reproductive Justice," 2005. Accessed August 6, 2016. http://strongfamiliesmovement.org/assets/docs/ACRJ-A-New-Vision.pdf.

Aziz, K. M. A., and W. Henry Mosley. "The History, Methodology and Main Findings of the Matlab Project in Bangladesh." Paper presented at IUSSP *Seminar on Socio-Cultural Determinants of Morbidity and Mortality in Developing Countries: The Role of Longitudinal Studies,* Saly Portudal, Senegal, October 9–11, 1991.

———. "The History, Methodology and Main Findings of the Matlab Project in Bangladesh." In *Prospective Community Studies in Developing Countries,* edited by Monica Das Gupta, Peter Aaby, Michael Garenne, and Gilles Pison, 28–53. Oxford: Clarendon, 1997.

Bandarage, Asoka. *Women, Population and Global Crisis: A Political-Economic Analysis*. London: Zed Books, 1997.

Barkan, Elazar. *The Retreat of Scientific Racism: Changing Concepts of Race in Britain and the United States between the World Wars*. Cambridge: Cambridge University Press, 1992.

Barrett, John C. "Potential Fertility and Averted Births." *Genus* 37, nos. 1/2 (1981): 1–15.

Barro, Robert J., and Gary S. Becker. "Fertility Choice in a Model of Economic Growth." *Econometrica* 57, no. 2 (1989): 481–501.

Bashford, Alison. *Global Population: History, Geopolitics, and Life on Earth*. New York: Columbia University Press, 2014.

Baucom, Ian. *Specters of the Atlantic: Finance Capital, Slavery, and the Philosophy of History*. Durham, NC: Duke University Press, 2005.

Becker, Gary S. "Investment in Human Capital: A Theoretical Analysis." *Journal of Political Economy* 70, no. 5 (1962): 9–49.

———. "On the Relevance of the New Economics of the Family." *American Economic Review* 64, no. 2 (1974): 317–19.

———. "A Theory of the Allocation of Time." *The Economic Journal* 75, no. 299 (1965): 493–517.

Becker, Gary S., and Robert J. Barro. "Altruism and the Economic Theory of Fertility." *Population and Development Review* 12 (1986):69–76.

———. "A Reformulation of the Economic Theory of Fertility." *Quarterly Journal of Economics* 103, no. 1 (1988): 1–25.

Becker, Gary S., and Nigel Tomes. "Child Endowments and the Quantity and Quality of Children." *Journal of Political Economy* 84, no. 4 (1976): S143–62.

Bellacasa, Maria Puig de la. "Making Time for Soil: Technoscientific Futurity and the Pace of Care." *Social Studies of Science* 45, no. 5 (2015): 691–716.

Benenson, A. S., P. R. Joseph, and R. O. Oseasohn. "Cholera Vaccine Field Trials in East Pakistan: 1. Reaction and Antigenicity Studies." *Bulletin of the World Health Organization* 38, no. 3 (1968): 347–57.

Benjamin, Walter. *The Arcades Project*. Edited by Rolf Tiedemann. Translated by Howard Eiland and Kevin McLaughlin. Cambridge, MA: Belknap Press of Harvard University Press, 2002.

Berelson, Bernard. "Beyond Family Planning." *Studies in Family Planning* 1, no. 38 (1969): 1–16.

Bergeron, Suzanne. *Fragments of Development: Nation, Gender, and the Space of Modernity*. Ann Arbor: University of Michigan Press, 2004.

Berlant, Lauren. "America, 'Fat,' the Fetus." *Boundary* 2 21, no. 3 (1994): 145–95.

———. *Cruel Optimism*. Durham, NC: Duke University Press, 2011.

Bernal, Victoria, and Inderpal Grewal. *Theorizing NGOs: States, Feminisms, and Neoliberalism*. Durham, NC: Duke University Press, 2014.

Bhandari, Labdhi Pat Raj. "Communications for Social Marketing: A Methodology for Developing Communication Appeals for Family Planning Programs." PhD diss., Columbia University, 1976.

Bharadwaj, Aditya. "Experimental Subjectification: The Pursuit of Human Embryonic Stem Cells in India." *Ethnos* 79, no. 1 (2014): 84–107.

Bhattacharya, S. K. "History of Development of Oral Rehydration Therapy." *Indian Journal of Public Health* 38, no. 2 (1994): 39–43.

Bhuiyan, Syeda Nurjahan, and Rokeya Begum. "Quinacrine Non-Surgical Female Sterilization in Bangladesh." *Contraception* 64, no. 5 (2001): 281–86.

Birla, Ritu. *Stages of Capital: Law, Culture, and Market Governance in Late Colonial India*. Durham, NC: Duke University Press, 2009.

Bledsoe, Caroline H. *Contingent Lives: Fertility, Time, and Aging in West Africa*. Chicago: University of Chicago Press, 2002.

Bonneuil, Christophe. "Development as Experiment: Science and State Building in Late Colonial and Postcolonial Africa, 1930–1970." *Osiris*, 2nd ser., 15 (January 2000): 258–81.

Bose, Sarmila. *Dead Reckoning: Memories of the 1971 Bangladesh War*. London: Hurst, 2011.

Boserup, Ester. *Population and Technological Change: A Study of Long-Term Trends*. Chicago: University of Chicago Press, 1981.

Bouk, Dan. *How Our Days Became Numbered: Risk and the Rise of the Statistical Individual*. Chicago: University of Chicago Press, 2015.

Boyle, Kirk. "The Four Fundamental Concepts of Slavoj Žižek's Psychoanalytic Marxism." *International Journal of Žižek Studies* 2, no. 1 (2008): 1–21.

BRAC. "Annual Report." Dhaka, Bangladesh: BRAC, 2013. Accessed April 4, 2015. https://www.brac.net/sites/default/files/annual-report-2013/BRAC-annual-report-2013.pdf.

Briggs, Laura. *Reproducing Empire: Race, Sex, Science, and U.S. Imperialism in Puerto Rico*. Berkeley: University of California Press, 2002.

Brookings Institution. "Janet Yellen's Dashboard." *Brookings*, June 10, 2014. Accessed March 13, 2016. https://www.brookings.edu/interactives/janet-yellens-dashboard/.

Browner, Carole H., and Carolyn F. Sargent, eds. *Reproduction, Globalization, and the State: New Theoretical and Ethnographic Perspectives*. Durham, NC: Duke University Press, 2011.

Buck-Morss, Susan. *The Dialectics of Seeing: Walter Benjamin and the Arcades Project*. Cambridge, MA: MIT Press, 1989.

Burki, Shahid Javed. "Twenty Years of the Civil Service of Pakistan: A Reevaluation." *Asian Survey* 9, no. 4 (1969): 239–54.

Caldwell, John C. "Demographers' Involvement in Twentieth-Century Population

Policy: Continuity or Discontinuity?" *Population Research and Policy Review* 24, no. 4 (2005): 359–85.

Çalışkan, Koray, and Michel Callon. "Economization, Part 1: Shifting Attention from the Economy towards Processes of Economization." *Economy and Society* 38, no. 3 (2009): 369–98.

———. "Economization, Part 2: A Research Programme for the Study of Markets." *Economy and Society* 39, no. 1 (2010): 1–32.

Callon, Michel, ed. *The Laws of the Markets*. Malden, MA: Blackwell, 1998.

Callon, Michel, Yuval Millo, and Fabian Muniesa, eds. *Market Devices*. Malden, MA: Blackwell, 2007.

Campbell, Chloe. *Race and Empire: Eugenics in Colonial Kenya*. Manchester: Manchester University Press, 2007.

Carson, Carol S. "The History of the United States National Income and Product Accounts: The Development of an Analytical Tool." *Review of Income and Wealth* 21, no. 2 (1975): 153–81.

Cash, Richard A. "A History of the Development of Oral Rehydration Therapy (ORT)." *Journal of Diarrhoeal Diseases Research* 5, no. 4 (1987): 256–61.

Casper, Monica J. *The Making of the Unborn Patient: A Social Anatomy of Fetal Surgery*. New Brunswick, NJ: Rutgers University Press, 1998.

Chaaban, Jad, and Wendy Cunningham. *Measuring the Economic Gain of Investing in Girls: The Girl Effect Dividend*. Policy Research Working Paper WPS5753. Washington, DC: World Bank, 2011.

Chakrabarty, Dipesh. "Climate and Capital: On Conjoined Histories." *Critical Inquiry* 41, no. 1 (2014): 1–23.

Chaia, Alberto, Aparna Dalal, Tony Goland, Maria Jose Gonzalez, Jonathan Morduch, and Robert Schiff. "Half the World Is Unbanked." Financial Access Initiative Framing Note, October 2009. Accessed December 7, 2013. http://mckinseyon society.com/half-the-world-is-unbanked/.

Chatterjee, Nilanjana, and Nancy E. Riley. "Planning an Indian Modernity: The Gendered Politics of Fertility Control." *Signs* 26, no. 3 (2001): 811–45.

Chatterjee, Partha. *The Politics of the Governed: Reflections on Popular Politics in Most of the World*. New York: Columbia University Press, 2004.

Chattopadhayay-Dutt, Purnima. *Loops and Roots: The Conflict between Official and Traditional Family Planning in India*. New Delhi: Ashish Publishing House, 1995.

Childs, Donald J. *Modernism and Eugenics: Woolf, Eliot, Yeats, and the Culture of Degeneration*. New York: Cambridge University Press, 2001.

Chowdhury, A. M. R., J. Patrick Vaughan, and F. H. Abed. "Use and Safety of Home-Made Oral Rehydration Solutions: An Epidemiological Evaluation from Bangladesh." *International Journal of Epidemiology* 17, no. 3 (1988): 655–65.

Christophers, Brett. "Making Finance Productive." *Economy and Society* 40, no. 1 (2011): 112–40.

Chua, Charmaine. "Logistics, Capitalist Circulation, Chokepoints." *The Disorder of Things* (blog), December 9, 2014. Accessed December 12, 2015. http://the disorderofthings.com/2014/09/09/logistics-capitalist-circulation-chokepoints/.

Clark, Colin. *The Conditions of Economic Progress*. London: Macmillan, 1940.

Clarke, Adele E. *Disciplining Reproduction: Modernity, American Life Sciences and the "Problems of Sex."* Berkeley: University of California Press, 1998.

Cleland, John, and W. Parker Mauldin. "The Promotion of Family Planning by Financial Payments: The Case of Bangladesh." *Studies in Family Planning* 22, no. 1 (1991): 1–18.

Clemens, John D. "Impact of B Subunity Killed Whole-Cell and Killed Whole-Cell Only Oral Vaccines against Cholera upon Treated Diarrhoeal Illness and Mortality in an Area Endemic for Cholera." *Lancet* 331, no. 8599 (1988): 1375–79.

Coale, Ansley J., and Edgar M. Hoover. "The Demographic Transition Reconsidered." In *International Population Conference, Liege*, 1:53–72. Liege: International Union for the Scientific Study of Population, 1973.

——. *Population Growth and Economic Development in Low-Income Countries: A Case Study of India's Prospects*. Princeton, NJ: Princeton University Press, 1958.

Coalition for Adolescent Girls. "The Most Powerful Person in the World" (2009). Accessed October 27, 2009, http://coalitionforadolescentgirls.org/.

Cohen, Lawrence. "Ethical Publicity: On Transplant Victims, Wounded Communities, and the Moral Demands of Dreaming." In *Ethical Life in South Asia*, edited by Anand Pandian and Daud Ali, 253–74. Bloomington: Indiana University Press, 2010.

——. "Operability, Bioavailability, and Exception." In *Global Assemblages: Technology, Politics, and Ethics as Anthropological Problems*, edited by Aihwa Ong and Stephen J. Collier, 79–90. Malden, MA: Blackwell, 2004.

Cohn, Bernard S. *Colonialism and Its Forms of Knowledge: The British in India*. Princeton, NJ: Princeton University Press, 1996.

Colen, Shellee. "'Like a Mother to Them': Stratified Reproduction and West Indian Childcare Workers and Employers in New York." In *Conceiving the New World Order: The Global Politics of Reproduction*, edited by Faye D. Ginsburg and Rayna Rapp, 78–102. Berkeley: University of California Press, 1995.

Connelly, Matthew. *Fatal Misconception: The Struggle to Control World Population*. Cambridge, MA: Belknap Press of Harvard University Press, 2008.

Cons, Jason, and Kasia Paprocki. "Contested Credit Landscapes: Microcredit, Self-Help and Self-Determination in Rural Bangladesh." *Third World Quarterly* 31, no. 4 (2010): 637–54.

Cooper, Frederick, and Randall Packard. "Introduction." In *International Development and the Social Sciences: Essays on the History and Politics of Knowledge*, edited by Frederick Cooper and Randall Packard, 1–41. Berkeley: University of California Press, 1997.

Cooper, Melinda. *Life as Surplus: Biotechnology and Capitalism in the Neoliberal Era.* Seattle: University of Washington Press, 2008.

———. "Resuscitations: Stem Cells and the Crisis of Old Age." *Body and Society* 12, no. 1 (2006): 1–23.

Cooper, Melinda, and Catherine Waldby. *Clinical Labor: Tissue Donors and Research Subjects in the Global Bioeconomy.* Durham, NC: Duke University Press, 2014.

Cordell, Dennis D., and Joel W. Gregory. *African Population and Capitalism: Historical Perspectives.* 2nd ed. Madison: University of Wisconsin Press, 1994.

Costa, Mariarosa Dalla, and Selma James. *The Power of Women and the Subversion of Community.* 3rd ed. Bristol, UK: Falling Wall Press, 1975.

Coughlin, William J. "Male Sterilization: Indian Festival." *Washington Post,* July 18, 1971, 1.

Cowen, Deborah. *The Deadly Life of Logistics: Mapping Violence in Global Trade.* Minneapolis: University of Minnesota Press, 2014.

Crichtlow, Donald T. *Intended Consequences: Birth Control, Abortion, and the Federal Government in Modern America.* Oxford: Oxford University Press, 1999.

Cristiano, Carlo. "Keynes and India, 1909–1913: A Study on Foreign Investment Policy." *European Journal of the History of Economic Thought* 16, no. 2 (2009): 301–24.

Curle, Adam. *Planning for Education in Pakistan: A Personal Case Study.* London: Tavistock Publications, 1966.

Curlin, George T., Lincoln C. Chen, and Sayed Babur Hussain. "Demographic Crisis: The Impact of the Bangladesh Civil War (1971) on Births and Deaths in a Rural Area of Bangladesh." *Population Studies* 30, no. 1 (1976): 87–105.

Curlin, G. T., R. J. Levine, A. Ahmed, K. M. A. Aziz, A. S. M. Rahman, and W. F. Verwey. *Immunological Aspects of a Cholera Toxoid Field Trial in Bangladesh.* Scientific Report 8. Dhaka: Cholera Research Laboratory, 1978.

Das, Maitreyi Bordia. *Whispers to Voices: Gender and Social Transformation in Bangladesh.* Washington, DC: World Bank, 2007.

Davie, Grace. *Poverty Knowledge in South Africa: A Social History of Human Science, 1855–2005.* New York: Cambridge University Press, 2015.

Davies, J., S. N. Mitra, and W. P. Schellstede. "Oral Contraception in Bangladesh: Social Marketing and the Importance of Husbands." *Studies in Family Planning* 18, no. 3 (1987): 157–68.

Davis, Angela Y. *Women, Race and Class.* New York: Vintage Books, 1983.

Davis, Kathy. *The Making of "Our Bodies, Ourselves": How Feminism Travels across Borders.* Durham, NC: Duke University Press, 2007.

Davis-Floyd, Robbie, and Joseph Dumit, eds. *Cyborg Babies: From Techno-Sex to Techno-Tots.* New York: Routledge, 1998.

Day, Sophie, Celia Lury, and Nina Wakeford. "Number Ecologies: Numbers and Numbering Practices." *Distinktion: Journal of Social Theory* 15, no. 2 (2014): 123–54.

D'Costa, Bina. *Nationbuilding, Gender, and War Crimes in South Asia*. New York: Routledge, 2011.

———. "Women, War, and the Making of Bangladesh: Remembering 1971." *Journal of Genocide Research* 14, no. 1(2012): 110–14.

D'Costa, Bina, and Sara Hossain. "Redress for Sexual Violence before the International Crimes Tribunal in Bangladesh: Lessons from History, and Hopes for the Future." *Criminal Law Forum* 21, no. 2 (2010): 331–59.

Dean, Mitchell M. *Governmentality: Power and Rule in Modern Society*. 2nd ed. Thousand Oaks, CA: SAGE Publications, 2009.

Demeny, Paul. "Observations on Population Policy and Population Program in Bangladesh." *Population and Development Review* 1, no. 2 (1975): 307–21.

Desrosières, Alain. *The Politics of Large Numbers: A History of Statistical Reasoning*. Translated by Camille Naish. Cambridge, MA: Harvard University Press, 2002.

Dhanraj, Deepa. *Something like a War*. New York: Women Make Movies. Documentary, 1991.

Donors' Community in Dhaka. "Position Paper on the Population Control and Family Planning Programme in Bangladesh." March 1, 1983.

Dubow, Saul. *Scientific Racism in Modern South Africa*. Cambridge: Cambridge University Press, 1995.

Dumit, Joseph. *Drugs for Life: How Pharmaceutical Companies Define Our Health*. Durham, NC: Duke University Press, 2012.

East, Edward M. "Food and Population." In *Proceedings of the World Population Conference, Geneva*, edited by Margaret Sanger, 85–91. London: Edward Arnold, 1927.

Edelstein, Michael. "The Size of the U.S. Armed Forces during World War II: Feasibility and War Planning." *Research in Economic History* 20 (November 2001): 47–97.

Ehrenreich, Barbara, and John Ehrenreich. *The American Health Empire: Power, Profits, and Politics*. Health/PAC Book. New York: Random House, 1971.

Ehrlich, Paul R. *The Population Bomb*. Rev. ed. New York: Ballantine Books, 1971.

Engels, Friedrich. *The Origin of the Family, Private Property, and the State: In the Light of the Researches of Lewis H. Morgan*. Translated by Ernest Untermann. Honolulu: University Press of the Pacific, 2001.

Engerman, David C., Nils Gilman, Mark H. Haefele, and Michael E. Latham, eds. *Staging Growth: Modernization, Development, and the Global Cold War*. Amherst: University of Massachusetts Press, 2003.

Enke, Stephen. "Birth Control for Economic Development." *Science* 164, no. 388 (1968): 798–802.

———. "The Economic Aspects of Slowing Population Growth." *Economic Journal* 76, no. 301 (1966): 44–56.

———. "The Economic Case for Birth Control." *Challenge* 15, no. 5 (1967): 30–43.

————. "An Economist Looks at Air Force Logistics." *Review of Economics and Statistics* 40, no. 3 (1958): 230–39.

————. "The Gains to India from Population Control: Some Money Measures and Incentive Schemes." *Review of Economics and Statistics* 42, no. 2 (1960): 175–81.

————. "Reducing Fertility to Accelerate Development." *Economic Journal* 84, no. 334 (1974): 349–66.

Erickson, Paul, Judy L. Klein, Lorraine Daston, Rebecca Lemov, Thomas Sturm, and Michael D. Gordin. *How Reason Almost Lost Its Mind: The Strange Career of Cold War Rationality*. Chicago: University of Chicago Press, 2013.

Fabian, Johannes. *Time and the Other: How Anthropology Makes Its Object*. New York: Columbia University Press, 2002.

Fairchild, Henry P. "Optimum Population." In *Proceedings of the World Population Conference, Geneva*, edited by Margaret Sanger, 72–84. London: Edward Arnold, 1927.

Fannin, Maria. "Domesticating Birth in the Hospital: 'Family-Centered' Birth and the Emergence of 'Homelike' Birthing Rooms." *Antipode* 35, no. 3 (2003): 513–35.

Fanon, Frantz. *Black Skin, White Masks*. Translated by Charles Lam Markmann. New York: Grove Press, 1967.

————. "Medicine and Colonialism." In *A Dying Colonialism*, translated by Haakon Chevalier, 121–46. New York: Grove Press, 1965.

————. "The 'North African Syndrome.'" In *Toward the African Revolution: Political Essays*, translated by Haakon Chevalier, 3–16. New York: Grove Press, 1967.

————. *The Wretched of the Earth*. New York: Grove Press, 1963.

Fassin, Didier. "Humanitarianism as a Politics of Life." *Public Culture* 19, no. 3 (2007): 499–520.

————. *Humanitarian Reason: A Moral History of the Present*. Berkeley: University of California Press, 2011.

Fauveau, Vincent. "Does ORT Reduce Diarrhoeal Mortality?" *Health Policy and Planning* 7, no. 3 (1992): 243–50.

————, ed. *Matlab: Women, Children and Health*. Dhaka: ICDDR,B, 1994.

Federici, Silvia. *Wages against Housework*. Bristol: Falling Wall Press, 1973.

Ferguson, James. *Global Shadows: Africa in the Neoliberal World Order*. Durham, NC: Duke University Press, 2006.

FHI 360. "Annual Report," 2013. Accessed April 9, 2015. http://www.fhi360.org/sites/default/files/media/documents/annual-report-2013.pdf.

FINRRAGE (Feminist International Network of Resistance to Reproductive and Genetic Engineering) and UBINIG. "Declaration of Comilla" (1991): vii–xi.

FINRRAGE-UBINIG International Conference, 1989 proceedings, held at BARD, Kotbari, Comilla, March 19–25.

Fisher, Jill A. *Medical Research for Hire: The Political Economy of Pharmaceutical Clinical Trials*. New Brunswick, NJ: Rutgers University Press, 2008.

Fixler, Dennis, and Bruce Grimm. "GDP Estimates: Rationality Tests and Turning Point Performance." *Journal of Productivity Analysis* 25, no. 3 (2006): 213–29.

Fornazzari, Alessandro. *Speculative Fictions: Chilean Culture, Economics, and the Neoliberal Transition.* Pittsburgh: University of Pittsburgh Press, 2013.

Fortun, Mike. *Promising Genomics: Iceland and deCODE Genetics in a World of Speculation.* Berkeley: University of California Press, 2008.

Foucault, Michel. *The Birth of Biopolitics: Lectures at the Collège de France, 1978–79.* Edited by Michel Senellart. Translated by Graham Burchell. New York: Palgrave Macmillan, 2008.

———. "Governmentality." In *The Foucault Effect: Studies in Governmentality*, edited by Graham Burchell, Colin Gordon, and Peter Miller, 87–104. Chicago: University of Chicago Press, 1991.

———. *History of Sexuality.* Vol. 1, *An Introduction.* New York: Random House, 1978.

———. *Security, Territory, Population: Lectures at the Collège de France, 1977–78.* Edited by Michel Senellart. Translated by Graham Burchell. New York: Palgrave Macmillan, 2007.

———. *Society Must Be Defended: Lectures at the Collège de France, 1975–76.* Edited by Mauro Bertani and Alessandro Fontana. Translated by David Macey. New York: Picador, 2003.

Franklin, Sarah. *Biological Relatives: IVF, Stem Cells, and the Future of Kinship.* Durham, NC: Duke University Press, 2013.

———. *Dolly Mixtures: The Remaking of Genealogy.* Durham, NC: Duke University Press, 2007.

———. "Ethical Biocapital: New Strategies of Cell Culture." In *Remaking Life and Death: Toward an Anthropology of the Biosciences*, edited by Sarah Franklin and Margaret Locke, 97–128. Santa Fe: School of American Research Press, 2003.

Franklin, Sarah, and Margaret Lock, eds. *Remaking Life and Death: Toward an Anthropology of the Biosciences.* Santa Fe: School of American Research Press, 2003.

Franklin, Sarah, and Celia Roberts. *Born and Made: An Ethnography of Preimplantation Genetic Diagnosis.* Princeton, NJ: Princeton University Press, 2006.

Fruhstuck, Sabine. *Colonizing Sex: Sexology and Social Control in Modern Japan.* Berkeley: University of California Press, 2003.

Fukuda-Parr, Sakiko. "Rescuing the Human Development Concept from the HDI: Reflections on a New Agenda." In *Readings in Human Development: Concepts, Measures and Policies for a Development Paradigm*, edited by Sakiko Fukuda-Parr and A. K. Shiva Kumar, 117–24. Oxford: Oxford University Press, 2003.

Gates, Melinda. "Bridging the Gender Gap: How Big Data Can Improve the Lives of a Billion Women and Girls." *Foreign Policy*, July 13, 2013.

Geissler, Paul Wenzel. *Para-States and Medical Science: Making African Global Health.* Durham, NC: Duke University Press, 2015.

Gilbert, Milton. "War Expenditures and National Production." *Survey of Current Business* 22, no. 3 (March 1942): 9–16.

Gilbert, Milton, and R. B. Bangs. "Preliminary Estimates of Gross National Product, 1929–41." *Survey of Current Business* 22, no. 5 (May 1942): 9–13.

Gilman, Nils. *Mandarins of the Future: Modernization Theory in Cold War America.* Baltimore: Johns Hopkins University Press, 2004.

Gilmore, Ruth Wilson. *Golden Gulag: Prisons, Surplus, Crisis, and Opposition in Globalizing California.* Berkeley: University of California Press, 2007.

Gil-Riaño, Sebastián. "Historicizing Anti-Racism: UNESCO's Campaigns against Race Prejudice in the 1950s." PhD diss., University of Toronto, 2014.

Ginsburg, Faye D., and Rayna Rapp, eds. *Conceiving the New World Order: The Global Politics of Reproduction.* Berkeley: University of California Press, 1995.

Girl Effect. "Assets and Insight: Invisible Girls." December 6, 2012. Accessed April 14, 2015. https://hubslide.com/girl-effect/assets-and-insight-invisible-girls-s56d71b788d070ead0e8c6513.html.

————. "Data and Development: Smarter Working." December 10, 2012. Accessed November 22, 2013. http://www.girleffect.org/what-girls-need/articles/2012/10/data-and-development-smarter-working.

————. "Economically Empowered Girls Can Stop Poverty before It Starts." January 31, 2013. Accessed November 5, 2013. http://www.girleffect.org/media/1163/girl_effect_infographic_economically-empowered.pdf.

————. "Empowering Girls with Economic Assets." January 31, 2013. Accessed August 6, 2016. http://www.girleffect.org/media/1187/girl_effect_presentation_empowering-girls-with-economic-assets.pdf.

————. "Empowering Girls with Economic Control." October 5, 2013. Accessed December 2, 2013. http://www.girleffect.org/what-girls-need/articles/2013/05/why-girls-need-economic-control/.

————. "Global Availability of Data on Girls." March 25, 2013. Accessed November 11, 2013. http://www.girleffect.org/explore/girl-data-making-the-numbers-add-up/data-invisble-girls/

————. "The State of Girl Data." November 10, 2012. Accessed April 18, 2015. http://www.girleffect.org/media/1178/girl_effect_feature_the-state-of-girl-data.pdf.

Girls Discovered. "Global Data Gaps," 2013. Accessed November 11, 2013. http://girlsdiscovered.org/media/v_girls_count-20131218_131931/girls_count/pdf/Girls_Discovered_Data_Gaps.pdf.

Gore, Timothy. "Extreme Carbon Inequality: Why the Paris Climate Deal Must Put the Poorest, Lowest Emitting and Most Vulnerable People First." Oxfam Media Briefing, December 2, 2015.

Goswami, Manu. *Producing India: From Colonial Economy to National Space.* Chicago: Chicago University Press, 2004.

Grameen Creative Lab. "Network." *Grameen Creative Lab*. (April 2015) Accessed September 28, 2016. http://www.grameencreativelab.com/our-company/network
.html.

Greenhalgh, Susan, and Edwin A. Winckler. *Governing China's Population: From Leninist to Neoliberal Biopolitics*. Stanford: Stanford University Press, 2005.

Grimes, David A., Herbert B. Peterson, Michael J. Rosenberg, John I. Fishburne, Roger W. Rochat, Atiqur R. Khan, and Rafiqul Islam. "Sterilization-Attributable Deaths in Bangladesh." *International Journal of Gynecology and Obstetrics* 20, no. 2 (1982): 149–54.

Grossbard-Shechtman, Shoshana. "The New Home Economics at Colombia and Chicago." *Feminist Economics* 7, no. 3 (November 2001): 103–30.

Grosz, Elizabeth. "Darwin and the Ontology of Life." In *Time Travels: Feminism, Nature, Power*, 35–42. Durham, NC: Duke University Press, 2005.

Guerrant, Richard L. "Cholera, Diarrhea, and Oral Rehydration Therapy: Triumph and Indictment." *Clinical Infectious Diseases* 37, no. 3 (2003): 398–405.

Gupta, Akhil. *Red Tape: Bureaucracy, Structural Violence, and Poverty in India*. Durham, NC: Duke University Press, 2012.

Guttmacher Institute. "2.3 Million Births Averted in Korea, Savings Huge." *International Family Planning Digest* 2, no. 2 (1976): 16.

Guyer, Jane. *Marginal Gains: Monetary Transactions in Atlantic Africa*. Chicago: University of Chicago Press, 2004.

Haider, Sheikh Kabir Uddin. "Genesis and Growth of the NGOs: Issues in Bangladesh Perspective." *International NGO Journal* 6, no. 11 (2011): 240–47.

Haraway, Donna J. "Anthropocene, Capitalocene, Plantationocene, Chthulucene: Making Kin." *Environmental Humanities* 6 (2015): 159–65.

———. *Modest_Witness@Second_Millennium: FemaleMan_Meets_OncoMouse*. New York: Routledge, 1997.

———. *When Species Meet*. Minneapolis: University of Minnesota Press, 2008.

———. "When Species Meet: Staying with the Trouble." *Environment and Planning D: Society and Space* 28, no. 1 (2010): 53–55.

Hardt, Michael, and Antonio Negri. *Empire*. Cambridge, MA: Harvard University Press, 2000.

Hartmann, Betsy. *Reproductive Rights and Wrongs: The Global Politics of Population Control*. Rev. ed. Boston: South End Press, 1995.

Hartmann, Betsy, and Hilary Standing. *Food, Saris and Sterilization: Population Control in Bangladesh*. London: Bangladesh International Action Group, 1985.

Harvey, Penny. *Roads: An Anthropology of Infrastructure and Expertise*. Ithaca, NY: Cornell University Press, 2015.

Harvey, Philip D. "Advertising Affordable Contraceptives: The Social Marketing Experience." In *Social Marketing: Theoretical and Practical Perspectives*, edited by Marvin E. Goldberg, Martin Fishbein, and Susan E. Middlestadt, 147–68. Mahwah, NJ: Lawrence Erlbaum Associates, 1997.

———. *Let Every Child Be Wanted: How Social Marketing Is Revolutionizing Contraceptive Use around the World.* Westport, CT: Auburn House, 1999.

Hasan, M. A. "Discovery of Numerous Mass Graves, Various Types of Torture on Women." October 18. Presentation, Bangladesh Study Group at Kean University, 2009.

Hasanat, Fayeza. "Sultana's Utopian Awakening: An Ecocritical Reading of Rokeya Sakhawat Hossain's *Sultana's Dream.*" *Asiatic* 7, no. 2 (2013): 114–25.

Hasnath, Syed Abu. "The Practice and Effect of Development Planning in Bangladesh: Summary." *Public Administration and Development* 7, no. 1 (1987): 59–75.

Hayden, Cori. *When Nature Goes Public: The Making and Unmaking of Bioprospecting in Mexico.* Princeton, NJ: Princeton University Press, 2003.

Hayhurst, Lyndsay M. C. "Corporatising Sport, Gender and Development: Postcolonial IR Feminisms, Transnational Private Governance and Global Corporate Social Engagement." *Third World Quarterly* 32, no. 3 (2011): 531–49.

Helmreich, Stefan. *Alien Ocean: Anthropological Voyages in Microbial Seas.* Berkeley: University of California Press, 2009.

———. "Blue-Green Capital, Biotechnological Circulation and an Oceanic Imaginary: A Critique of Biopolitical Economy." *BioSocieties* 2, no. 3 (2007): 287–302.

———. "Species of Biocapital." *Science as Culture* 17, no. 4 (2008): 463–78.

Hicks, John R. "Mr. Keynes and the 'Classics'; a Suggested Interpretation." *Econometrica* 5 (1937): 147–59.

Higgins, Patrick. *GDPNow: A Model for GDP "Nowcasting."* Working Paper Series 2014-7. Atlanta: Federal Reserve Bank of Atlanta, 2014.

Hodges, Sarah. *Contraception, Colonialism and Commerce: Birth Control in South India, 1920–1940.* Aldershot: Ashgate, 2008.

———. "Indian Eugenics in an Age of Reform." In *Reproductive Health in India: History, Politics, Controversies,* edited by Sarah Hodges, 115–38. New Delhi: Orient Longman, 2006.

———, ed. *Reproductive Health in India: History, Politics, Controversies.* New Delhi: Orient Longman, 2006.

Höhler, Sabine. "'Carrying Capacity'—The Moral Economy of the 'Coming Spaceship Earth.'" *Bilingual Journal of the Humanities and Social Sciences* 26, no. 1 (2006): 59–74.

———. *Spaceship Earth in the Environmental Age, 1960–1990.* London: Pickering and Chatto, 2014.

Hornblum, Allen M. *Acres of Skin: Human Experiments at Holmesburg Prison.* New York: Routledge, 1999.

Hossain, Yasmin. "The Begum's Dream: Rokeya Sakhawat Hossain and the Broadening of Muslim Women's Aspirations in Bengal." *South Asia Research* 12, no. 1 (1992): 1–19.

Hossein, Roquiah Sakhawat. *Sultana's Dream*. In *God Gives, Man Robs and Other Writings*, 31–47. Dhaka: Narigrantha Prabartana, 2002.

Hull, Matthew S. *Government of Paper: The Materiality of Bureaucracy in Urban Pakistan*. Berkeley: University of California Press, 2012.

Hustak, Carla. "Radical Intimacies: Affective Potential and the Politics of Love in the Transatlantic Sex Reform Movement, 1900–1930." PhD diss., University of Toronto, 2010.

Igo, Sarah E. *The Averaged American: Surveys, Citizens, and the Making of a Mass Public*. Cambridge, MA: Harvard University Press, 2008.

Innovative Projects in Family Planning and Rural Institutions in Bangladesh: Report on a National Workshop. Dhaka: Population Control and Family Planning Division, Government of the People's Republic of Bangladesh, 1978.

Islam, Kajalie Shehreen. "Breaking Down the *Birangona*: Examining the (Divided) Media Discourse on the War Heroines of Bangladesh's Independence Movement." *International Journal of Communication* 6 (2012): 2131–48.

Islam, Nurul. *Making of a Nation, Bangladesh: An Economist's Tale*. Dhaka: University Press, 2003.

Ittmann, Karl, Dennis D. Cordell, and Gregory H. Maddox, eds. *The Demographics of Empire: The Colonial Order and the Creation of Knowledge*. Athens: Ohio University Press, 2010.

James, Selma. *Sex, Race and Class—The Perspective of Winning: A Selection of Writings 1952–2011*. Oakland, CA: PM Press, 2012.

Janssens, Angélique. "'Were Women Present at the Demographic Transition?': A Question Revisited." *History of the Family* 12, no. 1 (2007): 43–49.

Johnson, Lyndon B. "Address in San Francisco at the 20th Anniversary Commemorative Session of the United Nations." June 25, 1965. Accessed August 6, 2016. http://www.presidency.ucsb.edu/ws/?pid=27054.

Jordanova, Ludmilla. "Interrogating the Concept of Reproduction in the Eighteenth Century." In *Conceiving the New World Order: The Global Politics of Reproduction*, edited by Faye D. Ginsburg and Rayna Rapp, 369–86. Berkeley: University of California Press, 1995.

Kaler, Amy. *Running after Pills: Gender, Politics and Contraception in Colonial Rhodesia*. Portsmouth, NH: Heinemann, 2003.

Kalpagam, U. "Colonial Governmentality and the 'Economy.'" *Economy and Society* 29, no. 3 (2000): 418–38.

———. *Rule by Numbers: Governmentality in Colonial India*. Lanham, MD: Lexington Books, 2014.

Kangas, Lenni W. "Integrated Incentives for Fertility Control." *Science*, n.s., 169, no. 3952 (1970): 1278–83.

Karim, Lamia. "Demystifying Micro-Credit: The Grameen Bank, NGOs, and Neoliberalism in Bangladesh." *Cultural Dynamics* 20, no. 1 (2008): 5–29.

———. *Microfinance and Its Discontents: Women in Debt in Bangladesh*. Minneapolis: University of Minnesota Press, 2011.

Kelly, Ann H., and P. Wenzel Geissler. *The Value of Transnational Medical Research: Labour, Participation and Care*. London: Routledge, 2013.

Kenny, Katherine E. "The Biopolitics of Global Health: Life and Death in Neoliberal Time." *Journal of Sociology* 51, no. 1 (2015): 9–27.

Keraan, Tauriq. "Banking the Bottom of the Pyramid" report, Deloitte, UK, 2010.

Kessel, E. "100,000 Quinacrine Sterilizations." *Advances in Contraception* 12, no. 2 (1996): 69–76.

Kevles, Daniel J. *In the Name of Eugenics: Genetics and the Uses of Human Heredity*. Berkeley: University of California Press, 1985.

Keyfitz, Nathan. "How Birth Control Affects Births." *Biodemography and Social Biology* 18, no. 2 (1971): 109–21.

Keynes, John Maynard. *The Economic Consequences of the Peace*. New York: Harcourt, Brace and Howe, 1920.

———. *Essays in Persuasion*. New York: W. W. Norton, 1931.

———. *The General Theory of Employment, Interest and Money*. London: Macmillan, 1936.

———. *Indian Currency and Finance*. London: Macmillan, 1913.

Khan, Ayesha. "Policy-Making in Pakistan's Population Programme." *Health Policy and Planning* 11, no. 1 (1996): 30–51.

Khan, Monirul Islam, Mushtaque Chowdury, and Abbas Bhuiya. "An Inventory of the Development Programmes by Government and Non-Government Organizations in Selected Unions of Matlab (Excluding BRAC–ICDDR,B)." Working Paper. Dhaka, Bangladesh: BRAC–ICDDR,B Joint Research Project, 1997.

Kingsland, Sharon E. *The Evolution of American Ecology, 1890–2000*. Baltimore: Johns Hopkins University Press, 2005.

Kligman, Gail. *The Politics of Duplicity: Controlling Reproduction in Ceausescu's Romania*. Berkeley: University of California Press, 1998.

Kline, Wendy. *Building a Better Race: Gender, Sexuality and Eugenics from the Turn of the Century to the Baby Boom*. Berkeley: University of California Press, 2001.

Kloppenburg, Jack Ralph, Jr. *First the Seed: The Political Economy of Plant Biotechnology*. 2nd ed. Madison: University of Wisconsin Press, 2005.

Koffman, Ofra, and Rosalind Gill. "'The Revolution Will Be Led by a 12-Year-Old Girl': Girl Power and Global Biopolitics." *Feminist Review* 105, no. 1 (2013): 83–102.

Kramer, Marcia. "Legal Abortion among New York City Residents: An Analysis According to Socioeconomic Demographic Characteristics." *Family Planning Perspectives* 7, no. 3 (1975): 128–37.

Kreager, Philip. "Population and the Making of the Human Sciences: A Historical Outline." In *Population in the Human Sciences: Concepts, Models, Evidence*, edited by Philip Kreager, Bruce Winney, Stanley Ulijaszek, and Cristian Capelli, 55–85. New York: Oxford University Press, 2015.

Krishnakumar, S. "Ernakulam's Third Vasectomy Campaign Using the Camp Approach." *Studies in Family Planning* 5, no. 2 (1974): 58–61.

———. "Kerala's Pioneering Experiment in Massive Vasectomy Camps." *Studies in Family Planning* 3, no. 8 (1972): 177–85.

Kristof, Nicholas. "Smart Girls vs. Bombs." *New York Times*, April 11, 2015.

Kuznets, Simon. "National Income, 1929–1932." Bulletin 49. New York: National Bureau of Economic Research, 1934.

Landecker, Hannah. *Culturing Life: When Cells Became Technologies*. Cambridge, MA: Harvard University Press, 2007.

Landefeld, J. Steven, Eugene P. Seskin, and Barbara M. Fraumeni. "Taking the Pulse of the Economy: Measuring GDP." *Journal of Economic Perspectives* 22, no. 2 (2008): 193–216.

Lang, Sabine. "The NGOization of Feminism." In *Transitions, Environments, Translations: Feminisms in International Politics*, edited by Joan W. Scott, Cora Kaplan, and Debra Keates, 101–20. New York: Routledge, 1997.

Latham, Michael E. *Modernization as Ideology: American Social Science and "Nation Building" in the Kennedy Era*. Chapel Hill: University of North Carolina Press, 2000.

———. *The Right Kind of Revolution: Modernization, Development, and U.S. Foreign Policy from the Cold War to the Present*. Ithaca, NY: Cornell University Press, 2011.

Lawson, Sandra. "Women Hold Up Half the Sky." Global Economics Paper 164, 2008. Accessed February 22, 2010. http://www.goldmansachs.com/our-thinking/investing-in-women/bios-pdfs/women-half-sky-pdf.pdf.

Lazzarato, Maurizio. *Signs and Machines: Capitalism and the Production of Subjectivity*. Translated by Joshua David Jordan. Los Angeles, CA: Semiotext(e), 2014.

Levine, Ruth, et al. *Girls Count: A Global Investment and Action Agenda*. Washington, DC: Center for Global Development, 2008.

Levy, Jonathan. *Freaks of Fortune: The Emerging World of Capitalism and Risk in America*. Cambridge, MA: Harvard University Press, 2012.

Li, Tania Murray. *The Will to Improve: Governmentality, Development, and the Practice of Politics*. Durham, NC: Duke University Press, 2007.

Lorde, Audre. *A Burst of Light: Essays*. Ithaca, NY: Firebrand Books, 1988.

Luthra, R. "Contraceptive Social Marketing in the Third World: A Case of Multiple Transfer." *International Communication Gazette* 47, no. 3 (1991): 159–76.

Mackinnon, A. "Were Women Present at the Demographic Transition? Questions from a Feminist Historian to Historical Demographers." *Gender and History* 7, no. 2 (1995): 222–40.

Mahalanabis, D., A. B. Choudhuri, N. G. Bagchi, A. K. Bhattacharya, and T. W. Simpson. "Oral Fluid Therapy of Cholera among Bangladesh Refugees. 1973." *Bulletin of the World Health Organization* 79, no. 5 (2001): 473–79.

Makdisi, Saree, Cesare Casarino, and Rebecca Karl, eds. *Marxism beyond Marxism*. New York: Routledge, 1995.

Malitz, David. "The Costs and Benefits of Title XX and Title XIX Family Planning Services in Texas." *Evaluation Review* 8, no. 4 (1984): 519–36.

Mamdani, Mahmood. *The Myth of Population Control: Family, Caste and Class in an Indian Village*. London: Monthly Review Press, 1973.

Mamo, Laura. *Queering Reproduction: Achieving Pregnancy in the Age of Technoscience*. Durham, NC: Duke University Press, 2007.

Mani, Lata. *Contentious Traditions: The Debate on Sati in Colonial India*. Berkeley: University of California Press, 1998.

Markowitz, Gerald, and David Rosner. *Lead Wars: The Politics of Science and the Fate of America's Children*. Berkeley: University of California Press, 2013.

Marx, Karl, *Capital*, Vol. 1, *A Critique of Political Economy*. Translated by Samuel Moore and Edward Aveling, edited by Frederick Engels, revised according to the 4th German edition by Ernest Untermann. Chicago: Charles H. Kerr and Co., 1909.

———. "Results of the Immediate Process of Production." In *Capital*, Vol. 1, *A Critique of Political Economy*. Translated by Ben Fowkes, appendix. New York: Vintage Books, 1977.

Mascarenhas, Anthony. *Bangladesh: A Legacy of Blood*. London: Arnold Overseas, 1986.

Masco, Joseph. *The Nuclear Borderlands: The Manhattan Project in Post–Cold War New Mexico*. Princeton, NJ: Princeton University Press, 2006.

———. "'Survival Is Your Business': Engineering Ruins and Affect in Nuclear America." *Cultural Anthropology* 23, no. 2 (2008): 361–98.

———. *The Theater of Operations: National Security Affect from the Cold War to the War on Terror*. Durham, NC: Duke University Press, 2014.

Massumi, Brian. "National Enterprise Emergency: Steps toward an Ecology of Powers." *Theory, Culture and Society* 26, no. 6 (2009): 153–85.

———. *The Power at the End of the Economy*. Durham, NC: Duke University Press, 2014.

Mauldin, W. Parker. "Births Averted by Family Planning Programs." *Studies in Family Planning* 1, no. 33 (1968): 1–7.

Mazhar, Farhad. "Nayakrishi Andolon." UBINIG, December 19, 2011. Accessed February 16, 2015. http://ubinig.org/index.php/home/showAerticle/32/english.

Mazhar, Farhad, P. V. Satheesh, Daniel Buckles, and Farida Akhter. "Uncultivated Food: The Missing Link in Livelihood and Poverty Programs." SANFEC Policy Brief 1. Dhaka: South Asia Network on Food, Ecology and Culture, 2002.

Mbembe, Achille. "Aesthetics of Superfluity." *Public Culture* 16, no. 3 (2004): 373–405.

———. "Necropolitics." *Public Culture* 15, no. 1 (2003): 11–40.

McCann, C. "Malthusian Men and Demographic Transitions: A Case Study of Hegemonic Masculinity in Mid-Twentieth-Century Population Theory." *Frontiers* 30, no. 1 (2009): 142–71.

McCarthy, James. "Contraceptive Sterilization in Four Latin American Countries." *Journal of Biosocial Science* 14, no. 2 (1982): 189–201.

McClintock, Anne. *Imperial Leather: Race, Gender, and Sexuality in the Colonial Contest.* New York: Routledge, 1995.

McKenzie, Donald, Fabian Muniesa, and Lucia Siu, eds. *Do Economists Make Markets? On the Performativity of Economics.* Princeton, NJ: Princeton University Press, 2007.

McKittrick, Katherine. "Mathematics Black Life." *Black Scholar* 44, no. 2 (2014): 16–28.

McLaren, Angus. *Our Own Master Race: Eugenics in Canada, 1885–1945.* Toronto, ON: McClelland and Stewart, 1990.

Meadows, Donella H., Dennis L. Meadows, Jorgen Randers, and William W. Behrens III. *The Limits to Growth: A Report for the Club of Rome's Project on the Predicament of Mankind.* New York: Universe Books, 1972.

Mehos, Donna, and Suzanne Moon. "The Uses of Portability: Circulating Experts in the Technopolitics of Cold War and Decolonization." In *Entangled Geographies: Empire and Technopolitics in the Global Cold War*, edited by Gabrielle Hecht, 43–74. Cambridge, MA: MIT Press, 2011.

Menen, Aubrey. "The Rapes of Bangladesh." *New York Times*, July 23, 1972.

Merchant, Emily R. "Prediction and Control: Global Population, Population Science, and Population Politics in the Twentieth Century." PhD diss., University of Michigan, 2015.

Mezzano, Michael. "The Progressive Origins of Eugenics Critics: Raymond Pearl, Herbert S. Jennings, and the Defense of Scientific Inquiry." *Journal of the Gilded Age and Progressive Era* 4, no. 1 (2005): 83–97.

Middleton, Drew. "Kennedy Is Widely Hailed in Foreign Capitals: Allies Warm Worries over U.S. Economy Damped." *New York Times*, February 5, 1961, Review of the Week Editorials.

Million, Dian. *Therapeutic Nations: Healing in an Age of Indigenous Human Rights.* 2nd ed. Tucson: University of Arizona Press, 2013.

Mills, Charles Wade. *The Racial Contract.* Ithaca, NY: Cornell University Press, 1997.

Mitchell, Michele. *Righteous Propagation: African Americans and the Politics of Racial Destiny after Reconstruction.* Chapel Hill: University of North Carolina Press, 2004.

Mitchell, Timothy. "Fixing the Economy." *Cultural Studies* 12, no. 1 (1998): 82–101.

———. *Rule of Experts: Egypt, Techno-Politics, Modernity.* Berkeley: University of California Press, 2002.

———. "The Work of Economics: How a Discipline Makes Its World." *European Journal of Sociology* 47, no. 2 (2005): 297–320.

Mitra-Kahn, Benjamin. "Redefining and Measuring the Economy since the Year 1600." PhD diss., City University London, 2010.

Moeller, Kathryn. "Proving 'The Girl Effect': Corporate Knowledge Production and Educational Intervention." *International Journal of Educational Development* 33, no. 6 (2013): 612–21.

Mohan, Manendra. *Advertising Management: Concepts and Cases*. New Delhi: Tata McGraw-Hill, 1989.

Mookherjee, Nayanika. "The Absent Piece of Skin: Gendered, Racialized and Territorial Inscriptions of Sexual Violence during the Bangladesh War." *Modern Asian Studies* 46, no. 6 (2012): 1572–1601.

———. "Available Motherhood: Legal Technologies, 'State of Exception' and the Dekinning of 'War Babies' in Bangladesh." *Childhood* 14, no. 3 (2007): 339–54.

———. "The 'Dead and Their Double Duties': Mourning, Melancholia, and the Martyred Intellectual Memorials in Bangladesh." *Space and Culture* 10, no. 2 (2007): 271–91.

———. "Gendered Embodiments: Mapping the Body-Politic of the Raped Woman and the Nation in Bangladesh." *Feminist Review* 88, no. 1 (2008): 36–53.

———. "'Remembering to Forget': Public Secrecy and Memory of Sexual Violence in the Bangladesh War of 1971." *Journal of the Royal Anthropological Institute* 12, no. 2 (2006): 433–50.

———. *The Spectral Wound: Sexual Violence, Public Memories, and the Bangladesh War of 1971*. Durham, NC: Duke University Press, 2015.

Moore, Hugh. *The Population Bomb*. New York: Population Policy Panel of the Hugh Moore Fund, 1959.

Moreland, R. Scott, Kong Chu, Edward Jacobson, and Thomas H. Naylor. "The Carolina Population Center Family Planning Administrator Training Game." *Management Science* 18, no. 12 (1972): B635–44.

Morgan, Jennifer. *Laboring Women: Reproduction and Gender in New World Slavery*. Philadelphia: University of Pennsylvania Press, 2004.

Morgan, Lynn M., and Elizabeth F. S. Roberts. "Reproductive Governance: Rights and Reproduction in Latin America." *Anthropology and Medicine* 19, no. 2 (2009): 241–54.

Morgan, Mary S. "Seeking Parts, Looking for Wholes." In *Histories of Scientific Observation*, edited by Lorraine Daston and Elizabeth Lunbeck, 303–25. Chicago: University of Chicago Press, 2011.

———. *The World in the Model: How Economists Work and Think*. New York: Cambridge University Press, 2012.

Morgan, Mary S., and Tarja Knuuttila. "Models and Modelling in Economics." SSRN Scholarly Paper ID 1499975. Rochester, NY: Social Science Research Network, 2008. Accessed March 31, 2011. http://papers.ssrn.com/abstract=1499975.

Morgen, Sandra. *Into Our Own Hands: The Women's Health Movement in the United States.* New Brunswick, NJ: Rutgers University Press, 2002.

Mosley, W. Henry, A. K. M. Alauddin Chowdhury, and K. M. A. Aziz. "Demographic Characteristics of a Population Laboratory in Rural East Pakistan." Center for Population Research, National Institute of Child Health and Human Development, 1970.

Mosley, Wiley H., William M. McCormack, M. Fahimuddin, K. M. A. Aziz, A. S. M. Mizanur Rahman, A. K. M. Alauddin Chowdhury, Albert R. Martin, John C. Feeley, and Robert A. Phillips. "Report of the 1966–67 Cholera Vaccine Field Trial in Rural East Pakistan: 1. Study Design and Results of the First Year of Observation." *Bulletin of the World Health Organization* 40, no. 2 (1969): 187–97.

Mosley, W. Henry, William E. Woodward, K. M. A. Aziz, A. S. M. Mizanur Rahman, A. K. M. Alauddin Chowdhury, Ansaruddin Ahmed, and John C. Feeley. "The 1968–1969 Cholera-Vaccine Field Trial in Rural East Pakistan: Effectiveness of Monovalent Ogawa and Inaba Vaccines and a Purified Inaba Antigen, with Comparative Results of Serological and Animal Protection Tests." *Journal of Infectious Diseases* 121 (May 1970): s1–9.

Mueller-Wille, Staffan. "Figures of Inheritance, 1650–1850." In *Heredity Produced: At the Crossroads of Biology, Politics, and Culture, 1500–1870*, edited by Staffan Mueller-Wille and Hans-Jörg Rheinberger, 177–204. Cambridge, MA: MIT Press, 2007.

Mukharji, Projit. "Swapnaushadhi: The Embedded Logic of Dreams and Medical Innovation in Bengal." *Culture, Medicine, and Psychiatry* 38, no. 3 (2014): 387–407.

Murphy, Michelle. "Distributed Reproduction." In *Corpus: An Interdisciplinary Reader on Bodies and Knowledge*, edited by Monica J. Casper and Paisley Currah. New York: Palgrave Macmillan, 2011.

———. "Distributed Reproduction, Chemical Violence, and Latency." *Scholar and Feminist Online* 11, no. 3 (2013). Accessed January 27, 2013. http://sfonline.barnard.edu/life-un-ltd-feminism-bioscience-race/distributed-reproduction-chemical-violence-and-latency/.

———. "Economization of Life: Calculative Infrastructures of Population and Economy." In *Relational Architectural Ecologies: Architecture, Nature and Subjectivity*, edited by Peg Rawes. London: Routledge, 2013.

———. "Reproduction." In *Marxism and Feminism*, edited by Shahrzad Mojab, 287–304. London: Zed Books, 2015.

———. *Seizing the Means of Reproduction: Entanglements of Feminism, Health, and Technoscience.* Durham, NC: Duke University Press, 2012.

Nalin, David R. 1968. "Oral Maintenance Therapy for Cholera in Adults." *Lancet* 292, no. 7564 (2015): 370–73.

Narigranth Prabartna. *On Shahaj Way to Ananda.* Bangladesh: Narigranth Prabartana, 1995.

National Academy of Sciences. *Rapid Population Growth: Consequences and Policy Implications*. Baltimore: Published for the National Academy of Sciences by the Johns Hopkins Press, 1971.

National Institute of Population Research and Training (NIPORT). "Bangladesh Demographic and Health Survey 2011." Dhaka, 2013.

National Security Council. *Implications of Worldwide Population Growth for U.S. Security and Overseas Interests*. National Security Study Memorandum 200. Washington, DC: National Security Council, 1974.

Nazar Soomro, Naureen, and Ghulam Murtaza Khoso. "Impact of Colonial Rule on Civil Services of Pakistan." *British Journal of Interdisciplinary Studies* 1, no. 2 (2014): 1–13.

Nazneen, Sohela, and Maheen Sultan. "Struggling for Survival and Autonomy: Impact of NGO-ization on Women's Organizations in Bangladesh." *Development* 52, no. 2 (2009): 193–99.

Nelson, Diane M. "Banal, Familiar, and Enrapturing: Financial Enchantment after Guatemala's Genocide." *Women's Studies Quarterly* 40, nos. 3/4 (2012): 205–25.

———. *Reckoning: The Ends of War in Guatemala*. Durham, NC: Duke University Press, 2009.

———. *Who Counts? The Mathematics of Death and Life after Genocide*. Durham, NC: Duke University Press, 2015.

Nelson, Jennifer. 2003. *Women of Color and the Reproductive Rights Movement*. New York: New York University Press.

Nguyen, Vinh-Kim. "Government-by-Exception: Enrolment and Experimentality in HIV Treatment Programmes in Africa." *Social Theory and Health* 7, no. 3 (2009): 196–217.

———. *The Republic of Therapy: Triage and Sovereignty in West Africa's Time of AIDS*. Durham, NC: Duke University Press, 2010.

Nike Inc. "Sustainable Business Performance Summary." Oregon, 2013.

Nortman, Dorothy. "Status of National Family Planning Programmes of Developing Countries in Relation to Demographic Targets." *Population Studies* 26, no. 1 (1972): 5–18.

Notestein, Frank W. "Abundant Life." In *Population Growth and Economic Development with Special Reference to Pakistan: Summary Report of a Seminar, September 8–13, 1959*, edited by M. L. Qureshi, 318–25. Karachi: Institute of Development Economics, 1960.

———. "Closing Remarks." In *Family Planning and Population Programs: A Review of World Developments*, edited by Bernard Berelson, 827–30. Chicago: University of Chicago Press, 1966.

———. "Population—The Long View." In *Food for the World*, edited by Theodore W. Schultz, 36–57. Chicago: University of Chicago Press, 1945.

Obermeyer, Ziad, Christopher J. L. Murray, and Emmanuela Gakidou. "Fifty Years of

Violent War Deaths from Vietnam to Bosnia: Analysis of Data from the World
Health Survey Programme." *British Medical Journal* 336, no. 7659 (June 26, 2008):
1482–86.

Orr, Jackie. *Panic Diaries: A Genealogy of Panic Disorder*. Durham, NC: Duke University
Press, 2006.

Osteria, T., Makhlisur Rahman, R. Langsten, Atiqur R. Khan, Douglas H. Huber, and
W. Henry Mosley. *Assessment of the Matlab Contraception Distribution Project:
Implications for Program Strategy*. Dhaka: Cholera Research Laboratory, 1978.

"Overpopulation, Overconsumption—In Pictures." *Guardian*, April 1, 2015.

Paddock, William, and Paul Paddock. *Famine—1975! America's Decision: Who Will Survive?* Boston: Little, Brown, 1967.

Pande, Amrita. *Wombs in Labor: Transnational Commercial Surrogacy in India*. New
York: Columbia University Press, 2014.

Papanek, Gustav F. *Pakistan's Development: Social Goals and Private Incentives*. Cambridge, MA: Harvard University Press, 1967.

Papanek, Gustav F., and Eshya Mujahid. "Effect of Policies on Agricultural Development: A Comparison of the Bengals and the Punjabs." *American Journal of Agricultural Economics* 65, no. 2 (1983): 426–37.

Park, Jin-kyung. "Bodies for Empire: Biopolitics, Reproduction, and Sexual Knowledge
in Late Colonial Korea." *Korean Journal of Medical History* 23, no. 2 (2014): 203–38.

Parry, Manon. *Broadcasting Birth Control: Mass Media and Family Planning*. New
Brunswick, NJ: Rutgers University Press, 2013.

Pasternak, Shiri. "The Economics of Insurgency: Thoughts on Idle No More and Critical Infrastructure." *Media Co-op*, January 14, 2013. Accessed December 12, 2015.
http://www.mediacoop.ca/story/economics-insurgency/15610.

———. "The Fiscal Body of Sovereignty: To 'Make Live' in Indian Country." *Settler
Colonial Studies* 6, no. 4 (October 1, 2016): 317–38.

———. "To 'Make Life' in Indian Country: Chief Theresa Spence and the Fiscal
Body of Settler Colonialism." Paper presented at the Intersections Lecture Series,
University of Toronto, March 5, 2015.

Patel, Geeta. "Advertisements, Proprietary Heterosexuality, and Hundis: Postcolonial
Finance, Nation-State Formations, and the New Idealized Family." *Rethinking
Marxism* 24, no. 4 (2012): 516–35.

———. "Risky Subjects: Insurance, Sexuality, and Capital." *Social Text* 24, no. 4
(2006): 25–65.

Pateman, Carole. *The Sexual Contract*. Stanford: Stanford University Press, 1988.

Paul, Diane B. *Controlling Human Heredity: 1865 to the Present*. Atlantic Highlands, NJ:
Humanity Press International, 1995.

Pearl, Raymond. "Biology of Population Growth." In *Proceedings of the World Population Conference, Geneva*, edited by Margaret Sanger, 22–38. London: Edward
Arnold, 1927.

————. *The Biology of Population Growth*. Rev. ed. New York: Alfred A. Knopf, 1930.

————. "Differential Fertility." *Quarterly Review of Biology* 2, no. 1 (1927): 102–18.

Pernick, Martin S. *The Black Stork: Eugenics and the Death of "Defective" Babies in American Medicine and Motion Pictures since 1915*. Oxford: Oxford University Press, 1999.

Peterson, Kristin. *Speculative Markets: Drug Circuits and Derivative Life in Nigeria*. Durham, NC: Duke University Press, 2014.

Petryna, Adriana. *When Experiments Travel: Clinical Trials and the Global Search for Human Subjects*. Princeton: Princeton University Press, 2009.

Phillips, James F., Wayne S. Stinson, Shushum Bhatia, Makhlisur Rahman, and J. Chakraborty. "The Demographic Impact of the Family Planning–Health Services Project in Matlab, Bangladesh." *Studies in Family Planning* 13, no. 5 (1982): 131–40.

Pierce, Nathaniel F. "How Much Has ORT Reduced Child Mortality?" *Journal of Health, Population and Nutrition* 19, no. 1 (2001): 1–3.

Plan International UK. "Girls in the Global Economy: Adding It All Up," 2009. Accessed April 2, 2013. http://www.plan-uk.org/assets/Documents/pdf/Because _I_am_a_Girl_2009_-_Executive_Summary.pdf.

Planning Commission. *The First Five Year Plan, 1973–78*. Dhaka: Government of the People's Republic of Bangladesh, 1973.

Population Control Programme in Bangladesh: Past, Present and Future. Dhaka: Ministry of Health and Population Control, Government of the People's Republic of Bangladesh, 1985.

Population Council. *A Manual for Surveys of Fertility and Family Planning: Knowledge, Attitudes, and Practice*. New York: Population Council, 1970.

————. "Taiwan: Births Averted by the IUD Program." *Studies in Family Planning* 1, no. 20 (1967): 7–8.

Porter, Theodore M. *Trust in Numbers: The Pursuit of Objectivity in Science and Public Life*. Princeton, NJ: Princeton University Press, 1995.

Potter, Robert G. "Additional Births Averted When Abortion Is Added to Contraception." *Studies in Family Planning* 3, no. 4 (1972): 53–59.

————. "Births Averted by Contraception: An Approach through Renewal Theory." *Theoretical Population Biology* 1, no. 3 (1970): 251–72.

Potts, Malcolm, and John M. Paxman. "Depo-Provera—Ethical Issues in Its Testing and Distribution." *Journal of Medical Ethics* 10, no. 1 (1984): 9–20.

Prakash, Gyan. *Another Reason: Science and the Imagination of Modern India*. Princeton, NJ: Princeton University Press, 1999.

PSI. "Progress Report." PSI, 2013. Accessed March 21, 2016. http://www.psi.org/wp -content/uploads/2014/10/PSI-Progress-Report-2013-English.pdf.

Quisumbing, Agnes, and Chiara Kovarik. *Investments in Adolescent Girls' Physical and Financial Assets*. New York: Nike Foundation; London: Department for International Development, 2013.

Raghavan, Srinath. *1971: A Global History of the Creation of Bangladesh*. Cambridge, MA: Harvard University Press, 2013.

Rahman, A. S. M. Mizanur, A. Majid Molla, Abdul Bari, and W. B. Greenough III. "Mothers Can Prepare and Use Rice-Salt Oral Rehydration Solution in Rural Bangladesh." *Lancet* 326, no. 8454 (1985): 539–40.

Rahman, Bazlur. "Expenditures and Funding of Population Programs in Bangladesh." Population, Development and Evaluation Unit, Family Health International, Research Triangle Park, NC, 1993.

Raman, Bhavani. *Document Raj: Writing and Scribes in Early Colonial South India*. Chicago: University of Chicago Press, 2012.

Ramsden, E. "Carving up Population Science: Eugenics, Demography and the Controversy over the 'Biological Law' of Population Growth." *Social Studies of Science* 32, nos. 5–6 (2002): 857–99.

Rankin, Katherine N. "A Critical Geography of Poverty Finance." *Third World Quarterly* 34, no. 4 (2013): 547–68.

———. "Social Capital, Microfinance, and the Politics of Development." *Feminist Economics* 8, no. 1 (2002): 1–24.

Rao, Mohan. *From Population Control to Reproductive Health: Malthusian Arithmetic*. New Delhi: SAGE Publications, 2004.

Rao, V. K. R. V. *The National Income of British India, 1931–1932*. London: Macmillan, 1940.

Rapp, Rayna. *Testing Women, Testing the Fetus: The Social Impact of Amniocentesis in America*. New York: Routledge, 2000.

Ravenholt, Reimert. "The Power of Availability." In *Village and Household Availability of Contraceptives: Africa/West Asia*, edited by J. S. Gardner, M. T. Mertaugh, M. Micklin, and G. Duncan. Seattle: Battelle Memorial Institute, Human Affairs Research Centers, 1977.

———. "World Fertility Survey." Conference paper presented at the World Fertility Survey Symposium, London, April 24, 1984.

Ravenholt, Reimert, and D. G. Gillespie. "Maximizing Availability of Contraception through Household Utilization." In *Village and Household Availability of Contraceptives: Southeast Asia*, edited by Jacqueline S. Gardner, R. Wolff, D. Gillespie, and G. Duncan. Seattle: Battelle Memorial Institute, Human Affairs Research Centers, 1976.

Ray, Bharati. *Women of India: Colonial and Post-Colonial Periods*. New Delhi: SAGE Publications, 2005.

Razzaque, Abdur. "Sociodemographic Differentials in Mortality during the 1974–75 Famine in a Rural Area of Bangladesh." *Journal of Biosocial Science* 21, no. 1 (1989): 13–22.

Razzaque, Abdur, Nurul Alam, Lokky Wai, and Andrew Foster. "Sustained Effects of the 1974–5 Famine on Infant and Child Mortality in a Rural Area of Bangladesh." *Population Studies* 44, no. 1 (1990): 145–54.

Redfield, Peter. "Bioexpectations: Life Technologies as Humanitarian Goods." *Public Culture* 24, no. 1 (2012): 157–84.

———. *Life in Crisis: The Ethical Journey of Doctors Without Borders*. Berkeley: University of California Press, 2013.

Reilly, Benjamin. "Great Bhola Cyclone, 1970." In *Disaster and Human History: Case Studies in Nature, Society and Catastrophe*, 172–84. Jefferson, NC: McFarland, 2009.

Reverby, Susan M. *Examining Tuskegee: The Infamous Syphilis Study and Its Legacy*. Chapel Hill: University of North Carolina Press, 2009.

Rheinberger, Hans-Jörg. "Experimental Systems: Historicality, Narration, and Deconstruction." *Science in Context* 7, no. 1 (1994): 65–81.

———. "Infra-Experimentality: From Traces to Data, from Data to Patterning Facts." *History of Science* 49, no. 164 (2011): 337–48.

———. *Toward a History of Epistemic Things: Synthesizing Proteins in the Test Tube*. Stanford: Stanford University Press, 1997.

R. L. "Wanderings of the Slave: Black Life and Social Death." *Mute*, June 5, 2013. Accessed April 15, 2015. http://www.metamute.org/editorial/articles/wanderings -slave-black-life-and-social-death.

Roberts, Celia. *Messengers of Sex: Hormones, Biomedicine, and Feminism*. New York: Cambridge University Press, 2007.

Roberts, Dorothy. *Killing the Black Body: Race, Reproduction, and the Meaning of Liberty*. New York: Pantheon Books, 1997.

Robertson, Thomas. *The Malthusian Moment: Global Population Growth and the Birth of American Environmentalism*. New Brunswick, NJ: Rutgers University Press, 2012.

Robinson, Warren C. "Family Planning in Pakistan 1955–1977: A Review." *Pakistan Development Review* 17, no. 2 (1978): 233–47.

Robinson, Warren C., Makhdoom A. Shah, and Nasra M. Shah. "The Family Planning Program in Pakistan: What Went Wrong?" *International Family Planning Perspectives* 7, no. 3 (1981): 85–92.

Roger, Jacques. *Buffon: A Life in Natural History*. Translated by Sarah Bonnefoi. Ithaca, NY: Cornell University Press, 1997.

Rogers, Everett M. *Communication Strategies for Family Planning*. New York: Free Press, 1973.

———. "Incentives in the Diffusion of Family Planning Innovations." *Studies in Family Planning* 2, no. 12 (1971): 241–48.

Rose, Nikolas. "Governing 'Advanced' Liberal Democracies." In *Foucault and Political Reason: Liberalism, Neo-Liberalism, and Rationalities of Government*, edited by Andrew Barry, Thomas Osbourne, and Nikolas Rose, 37–64. Chicago: University of Chicago Press, 1996.

———. *The Politics of Life Itself: Biomedicine, Power, and Subjectivity in the Twenty-First Century*. Princeton, NJ: Princeton University Press, 2007.

Rostow, Walt Whitman. *The Stages of Economic Growth: A Non-Communist Manifesto.* New York: Cambridge University Press, 1990.

Roy, Ananya. *Poverty Capital: Microfinance and the Making of Development.* New York: Routledge, 2010.

———. "Subjects of Risk: Technologies of Gender in the Making of Millennial Modernity." *Public Culture* 24, no. 1 66 (2012): 131–55.

Rudd, Peter. "The United Farm Workers Clinic in Delano, Calif.: A Study of the Rural Poor." *Rural Health* 90, no. 4 (1975): 331–39.

Ruxin, Joshua Nalibow. "Magic Bullet: The History of Oral Rehydration Therapy." *Medical History* 38, no. 4 (1994): 363.

Sack, David. "The Pak-SEATO Cholera Research Laboratory." Interview with W. H. Mosley. *Glimpse* 25, nos. 3–4 (2003): 5–7.

Sack, Jean, and M. A. Rahim, eds. *Smriti: ICDDR,B in Memory.* Dhaka: ICDDR,B, Centre for Health and Population Research, 2003.

Saikia, Yasmin. *Women, War, and the Making of Bangladesh: Remembering 1971.* Durham, NC: Duke University Press, 2011.

Samuelson, Paul Anthony. *Economics: An Introductory Analysis.* McGraw-Hill, 1948.

Samuelson, Paul Anthony, and William D. Nordhaus. *Economics.* 15th ed. New York: McGraw-Hill, 1995.

Sanger, Margaret, ed. *Proceedings of the World Population Conference, Geneva.* London: Edward Arnold, 1927.

Sarkar, Mahua. *Visible Histories, Disappearing Women: Producing Muslim Womanhood in Late Colonial Bengal.* Durham, NC: Duke University Press, 2008.

Saunders, Lyle. "Research and Evaluation: Needs for the Future." In *Family Planning and Population Programs: A Review of World Developments,* edited by Bernard Berelson. Chicago: University of Chicago Press, 1966.

Schabas, Margaret. *The Natural Origins of Economics.* Chicago: University of Chicago Press, 2005.

Schellstede, William P., and Robert L. Ciszewski. "Social Marketing of Contraceptives in Bangladesh." *Studies in Family Planning* 15, no. 1 (1984): 30–39.

Schultz, Stanley G. "From a Pump Handle to Oral Rehydration Therapy: A Model of Translational Research." *Advances in Physiology Education* 31, no. 4 (2007): 288–93.

Schultz, Theodore W. "The Economics of Being Poor: Noble Prize Lecture." Stockholm, December 8, 1979. http://nobelprize.org/nobel_prizes/economics/laureates/1979/schultz-lecture.html.

Schwittay, Anke. "The Financial Inclusion Assemblage: Subjects, Technics, Rationalities." *Critique of Anthropology* 31, no. 4 (December 1, 2011): 381–401.

Scott, David. "Colonial Governmentality." *Social Text* 43 (autumn 1995): 191–220.

———. *Conscripts of Modernity: The Tragedy of Colonial Enlightenment.* Durham, NC: Duke University Press, 2004.

Scully, Judith A. M. "Maternal Mortality, Population Control, and the War in Women's Wombs: A Bioethical Analysis of Quinacrine Sterilizations." *Wisconsin International Law Journal* 19, no. 2 (2001): 103–51.

Sebsted, Jennifer. "Girls and Their Money." Report, Nike Foundation and Microfinance Opportunities, November 2011. Accessed November 23, 2013. http://www.youtheconomicopportunities.org/sites/default/files/uploads/resource/Nike%20Report%202011.pdf.

Sen, Amartya. *Poverty and Famines: An Essay on Entitlement and Deprivation.* New York: Oxford University Press, 1981.

Setlur, Shivrang. "Searching for South Asian Intelligence: Psychometry in British India, 1919–1940." *Journal of the History of the Behavioral Sciences* 50, no. 4 (2014): 359–75.

Sexton, Jared. "The Social Life of Social Death: On Afro-Pessimism and Black Optimism." *InTensions* 5 (fall/winter 2011).

Silliman, Jael. "Expanding Civil Society: Shrinking Political Spaces: The Case of Women's NGOs." In *Dangerous Intersections: Feminist Perspectives on Population, Environment, and Development,* edited by Jael Silliman and Ynestra King, 23–53. Cambridge, MA: South End Press, 1998.

Silliman, Jael, and Anannya Bhattacharjee, eds. *Policing the National Body: Sex, Race, and Criminalization.* Cambridge, MA: South End Press, 2002.

Silliman, Jael, Marlene Gerber Fried, Loretta Ross, and Elena R. Gutiérrez. *Undivided Rights: Women of Color Organize for Reproductive Justice.* Cambridge, MA: South End Press, 2004.

Silva, Patricio. "Technocrats and Politics in Chile: From the Chicago Boys to the CIEPLAN Monks." *Journal of Latin American Studies* 23, no. 2 (1991): 385–410.

Simon, Julian L. "A Huge Marketing Research Task: Birth Control." *Journal of Marketing Research* 5, no. 1 (1968): 21–27.

Simone, Abdou Maliq. "People as Infrastructure: Intersecting Fragments in Johannesburg." *Public Culture* 16, no. 3 (2004): 407–29.

Simpson, Audra. *Mohawk Interruptus: Political Life across the Borders of Settler States.* Durham, NC: Duke University Press, 2014.

Simpson, Leanne. *Islands of Decolonial Love.* Winnipeg: Arbeiter Ring, 2013.

Sinha, Mrinalini. *Specters of Mother India: The Global Restructuring of an Empire.* Durham, NC: Duke University Press, 2006.

SisterSong Women of Color Reproductive Health Collective. *Reproductive Justice Briefing Book: A Primer on Reproductive Justice and Social Change.* Atlanta: SisterSong, 2007.

Sloterdijk, Peter. *Terror from the Air.* Translated by Amy Patton and Steve Corcoran. Los Angeles: Semiotext(e), 2009.

Smiley, Tavis. Interview of Secretary of State Hillary Rodham Clinton on Tavis Smiley

Reports. Television, January 27, 2010. Accessed September 24, 2016, http://www
.state.gov/secretary/20092013clinton/rm/2010/01/136489.htm.

Smilie, Ian. *Freedom from Want: The Remarkable Success Story of BRAC, the Global Grass-roots Organization That's Winning the Fight against Poverty.* Sterling, VA: Kumarian Press, 2009.

"Snapshot: Consumer Insights for Adolescent Girls — The Case of BRAC in Bangladesh." Groupe Speciale Mobile Association, 2014.

Sobhan, Rehman. "Politics of Food and Famine in Bangladesh." *Economic and Political Weekly* 14, no. 48 (1979): 1973–80.

Solinger, Rickie. *Beggars and Choosers: How the Politics of Choice Shapes Adoption, Abortion, and Welfare in the United States.* New York: Hill and Wang, 2002.

Speich, Daniel. "The Kenyan Style of 'African Socialism': Developmental Knowledge Claims and the Explanatory Limits of the Cold War." *Diplomatic History* 33, no. 3 (2009): 449–66.

———. "Traveling with the GDP through Early Development Economics' History." *The Nature of Evidence: How Well Do Facts Travel?* Working Paper 33/08, 2008. Accessed March 5, 2013. http://papers.ssrn.com/sol3/papers.cfm?abstract_id=1291058.

———. "The Use of Global Abstractions: National Income Accounting in the Period of Imperial Decline." *Journal of Global History* 6, no. 1 (2011): 7–28.

Star, Susan Lee. "The Ethnography of Infrastructure." *American Behavioral Scientist* 43, no. 3 (1999): 377–91.

Star, Susan Lee, and Karen Ruhleder. "Steps Towards an Ecology of Infrastructure." *Information Systems Research* 7, no. 1 (1996): 111–34.

Stengers, Isabella, Phillipe Pignarre, and Andrew Goffey. *Capitalist Sorcery: Breaking the Spell.* New York: Palgrave Macmillan, 2011.

Stepan, Nancy Leys. *"The Hour of Eugenics": Race, Gender, and Nation in Latin America.* Ithaca, NY: Cornell University Press, 1991.

Stern, Alexandra Minna. *Eugenic Nation: Faults and Frontiers of Better Breeding in Modern America.* Berkeley: University of California Press, 2005.

Stix, Regine K., and Frank W. Notestein. *Controlled Fertility: An Evaluation of Clinic Service.* Baltimore: Williams and Wilkins, 1940.

Stoler, Ann Laura. *Race and the Education of Desire: Foucault's "History of Sexuality" and the Colonial Order of Things.* Durham, NC: Duke University Press, 1995.

Strathern, Marilyn. *Reproducing the Future: Essays on Anthropology, Kinship, and the New Reproductive Technologies.* Manchester: Manchester University Press, 1992.

Street, Alice. *Biomedicine in an Unstable Place: Infrastructure and Personhood in a Papua New Guinean Hospital.* Durham, NC: Duke University Press, 2014.

Stycos, J. Mayone. "Survey Research and Population Control in Latin America." *Public Opinion Quarterly* 28, no. 3 (1964): 367–72.

Summers, Lawrence H. "Investing in All the People." World Bank, May 31, 1992.

http://documents.worldbank.org/curated/en/434361468739196605/Investing
-in-all-the-people.

Sunder Rajan, Kaushik. *Biocapital: The Constitution of Postgenomic Life*. Durham, NC: Duke University Press, 2006.

Suzuki, Tomo. "The Epistemology of Macroeconomic Reality: The Keynesian Revolution from an Accounting Point of View." *Accounting, Organizations and Society* 28, no. 5 (2003): 471–517.

Switzer, Heather. "(Post)Feminist Development Fables: The Girl Effect and the Production of Sexual Subjects." *Feminist Theory* 14, no. 3 (2013): 345–60.

Szreter, Simon. "The Idea of Demographic Transition and the Study of Fertility: A Critical Intellectual History." *Population and Development Review* 19, no. 4 (1993): 659–701.

———. "Theories and Heuristics: How Best to Approach the Study of Historic Fertility Declines?" *Historical Social Research/Historische Sozialforschung* 36, no. 2 (136) (2011): 65–98.

Szreter, Simon, Hania Sholkamy, and A. Dharmalingam, eds. *Categories and Contexts: Anthropological and Historical Studies in Critical Demography*. New York: Oxford University Press, 2004.

Tadiar, Neferti X. M. "Life-Times in Fate Playing." *South Atlantic Quarterly* 111, no. 4 (2012): 783–802.

Takeshita, Chikako. *The Global Biopolitics of the IUD: How Science Constructs Contraceptive Users and Women's Bodies*. Cambridge, MA: MIT Press, 2011.

Tallbear, Kim. "Why Interspecies Thinking Needs Indigenous Standpoints." Theorizing the Contemporary, *Cultural Anthropology* website, April 24, 2011. Accessed September 7, 2015. http://culanth.org/fieldsights/260-why-interspecies-thinking -needs-indigenous-standpoints.

Tan-Torres Edejer, Tessa, Moses Aikins, Robert Black, Lara Wolfson, Raymond Hutubessy, and David B. Evans. "Achieving the Millennium Development Goals for Health: Cost Effectiveness Analysis of Strategies for Child Health in Developing Countries." *British Medical Journal* 331, no. 7526 (2005): 1177–80.

Tarlo, Emma. *Unsettling Memories: Narratives of the Emergency in Delhi*. Berkeley: University of California Press, 2003.

Taussig, Karen-Sue, Rayna Rapp, and Deborah Heath. "Flexible Eugenics: Technologies of the Self in the Age of Genetics." In *Genetic Nature/Culture: Anthropology and Science beyond the Two-Culture Divide*, edited by Alan H. Goodman, Deborah Heath, and M. Susan Lindee, 51–76. Berkeley: University of California Press, 2003.

Taylor, C. E., and W. B. Greenough III. "Control of Diarrheal Diseases." *Annual Review of Public Health* 10, no. 1 (1989): 221–44.

Thompson, Charis. *Making Parents: The Ontological Choreography of Reproductive Technologies*. Cambridge, MA: MIT Press, 2005.

Thompson, Malcolm. "Foucault, Fields of Governability, and the Population–Family–Economy Nexus in China." *History and Theory* 51, no. 1 (2012): 42–62.

Ticktin, Miriam Iris. *Casualties of Care: Immigration and the Politics of Humanitarianism in France*. Berkeley: University of California Press, 2011.

Tilley, Helen. *Africa as a Living Laboratory: Empire, Development, and the Problem of Scientific Knowledge, 1870–1950*. Chicago: University of Chicago Press, 2011.

Tousignant, Noémi. "Insects-as-Infrastructure: Indicating, Project Locustox and the Sahelization of Ecotoxicology." *Science as Culture* 22, no. 1 (2013): 108–31.

Towghi, Fouzieyha. "Normalizing Off-Label Experiments and the Pharmaceuticalization of Homebirths in Pakistan." *Ethnos* 79, no. 1 (2014): 108–37.

Towghi, Fouzieyha, and Kalindi Vora. "Bodies, Markets, and the Experimental in South Asia." *Ethnos* 79, no. 1 (2014): 1–18.

Toye, John. *Keynes on Population*. New York: Oxford University Press, 2000.

Trumbull, Robert. "Dacca Raising the Status of Women While Aiding Rape Victims." *New York Times*, May 12, 1972.

Tuck, Eve, and K. Wayne Yang. "R-Words: Refusing Research." In *Humanizing Research: Decolonizing Qualitative Inquiry with Youth and Communities*, edited by Django Paris and Maisha T. Winn, 223–48. Thousand Oaks, CA: SAGE Publications, 2013.

UBINIG (Unnayan Bikalper Niti Nirdharoni Gobeshona). "About UBINIG." May 10, 2003. Accessed April 16, 2015. http://www.ubinig.org/index.php/campaigndetails/showAerticle/5/12.

———. *Faces of Coercion: Sterilization Tearing Apart Organs*. Dhaka: Narigrantha Prabartana, 1995.

———. *Violence of Population Control*. Dhaka: Narigrantha Prabartana, 1999.

———. *Women and Children of Bangladesh as Experimental Animals*. Dhaka: Narigrantha Prabartana, 1996.

Ul Haq, Mahbub. *Reflections on Human Development*. Oxford: Oxford University Press, 1995.

UNICEF. "The Unmet Need for Family Planning." *The Progress of Nations*. Wallingford, UK, 1995.

United Nations High Commissioner for Refugees (UNHCR). "The State of the World's Refugees 2000: Fifty Years of Humanitarian Action," 2000. Accessed April 8, 2015. http://www.unhcr.org/4a4c754a9.html.

USAID. "Disaster Relief: Bangladesh, Civil Strife, January–September, 1972." Foreign Disaster Case Report. Washington, DC: USAID, 1972.

USAID, Bangladesh. "Project Paper: Population/Family Planning, Bangladesh." Paper 388-0001. USAID, Dhaka, December 12, 1975.

U.S. Department of Commerce. "GDP: One of the Great Inventions of the 20th Century." *Survey of Current Business* 80, no. 1 (2000): 6–8.

———. "A Guide to the National Income and Product Accounts of the United

States," 2006. Accessed February 28, 2013. http://www.bea.gov/national/pdf /nipaguid.pdf.

van Ginneken, J., R. Bairagi, A. de Francisco, A. M. Sarder, and P. Vaughan. *Health and Demographic Surveillance in Matlab: Past, Present and Future.* Dhaka: ICDDR,B, 1998.

Van Hollen, Cecilia. *Birth in the Age of AIDS: Women, Reproduction, and HIV/AIDS in India.* Stanford: Stanford University Press, 2013.

―――. *Birth on the Threshold: Childbirth and Modernity in South India.* Berkeley: University of California Press, 2003.

Vanoli, André. *A History of National Accounting.* Amsterdam: IOS Press, 2005.

Varley, R. C., J. Tarvid, and D. N. Chao. "A Reassessment of the Cost-Effectiveness of Water and Sanitation Interventions in Programmes for Controlling Childhood Diarrhoea." *Bulletin of the World Health Organization* 76, no. 6 (1998): 617–31.

Venkatacharya, K., and N. P. Das. "An Application of a Monte Carlo Model to Estimate Births Averted Due to Various Family Planning Methods." *Sankhyā: The Indian Journal of Statistics, Series B* 34, no. 3 (1972): 297–310.

Verran, Helen. "Number as an Inventive Frontier in Knowing and Working Australia's Water Resources." *Anthropological Theory* 10, nos. 1–2 (2010): 171–78.

Vogler, Roger. "The Birth of Bangladesh: Nefarious Plots and Cold War Sideshows." *Pakistaniaat: A Journal of Pakistan Studies* 2, no. 3 (September 14, 2010): 24–46.

Vora, Kalindi. "Indian Transnational Surrogacy and the Commodification of Vital Energy." *Subjectivity* 28, no. 1 (2009): 266–78.

―――. *Life Support: Biocapital and the New History of Outsourced Labor.* Minneapolis: University of Minnesota Press, 2015.

Waldby, Catherine. "Stem Cells, Tissue Cultures and the Production of Biovalue." *Health* 6, no. 3 (2002): 305–23.

Waldby, Catherine, and Melinda Cooper. "The Biopolitics of Reproduction: Post-Fordist Biotechnology and Women's Clinical Labour." *Australian Feminist Studies* 23, no. 55 (2008): 57–73.

―――. "From Reproductive Work to Regenerative Labour: The Female Body and the Stem Cell Industries." *Feminist Theory* 11, no. 1 (2010): 3–22.

Waldby, Catherine, and Robert Mitchell. *Tissue Economies: Blood, Organs and Cell Lines in Late Capitalism.* Durham, NC: Duke University Press, 2006.

Walle, Etienne van de. "Fertility Transition, Conscious Choice, and Numeracy." *Demography* 29, no. 4 (1992): 487–502.

Walsh, B. Thomas. "Evaluation of the GE-TEMPO Project." American Public Health Association and USAID, 1975. Accessed August 1, 2013. http://pdf.usaid.gov/pdf _docs/Pdaaa497a1.pdf.

Waring, Marilyn. *If Women Counted: A New Feminist Economics.* New York: Harper-Collins, 1990.

Warwick, Donald P. "The KAP Survey: Dictates of Mission versus Demands of Science." In *Social Research in Developing Countries: Survey and Censuses in the Third World*, edited by Martin Bulmer and Donald P. Warwick, 349–64. New York: John Wiley and Sons, 1993.

Washington, Harriet A. *Medical Apartheid: The Dark History of Medical Experimentation on Black Americans from Colonial Times to the Present*. New York: Anchor, 2008.

Watkins, Susan Cotts. "If All We Knew about Women Was What We Read in *Demography*, What Would We Know?" *Demography* 30, no. 4 (1993): 551–77.

Werle, Nicholas. "More Than a Sum of Its Parts: A Keynesian Epistemology of Statistics." *Journal of Philosophical Economics* 4, no. 2 (2011): 65–92.

Widmer, Alexandra. "Of Temporal Politics and Demographic Anxieties: 'Young Mothers' in Demographic Predictions and Social Life in Vanuatu." *Anthropologica* 55, no. 2 (2013): 317–28.

———. "Unsettling Time." Paper presented at the Technoscience Salon, Toronto, December 3, 2015.

Williams, Rebecca. "Rockefeller Foundation Support to the Khanna Study: Population Policy and the Construction of Demographic Knowledge, 1945–1953." Rockefeller Archives Research Report, 2011.

World Bank. *World Development Indicators*. Washington, DC: World Bank, 2012. Accessed December 14, 2015. http://hdl.handle.net/10986/6014.

Wright, Melissa. *Disposable Women and Other Myths of Global Capitalism*. New York: Routledge, 2006.

Wyon, John B., and John E. Gordon. *The Khanna Study: Population Problems in the Rural Punjab*. Cambridge, MA: Harvard University Press, 1971.

Yunus, Muhammad. "The Problem of Poverty in Bangladesh." July 7, 2007. Accessed April 2, 2015. http://www.muhammadyunus.org/index.php/media/articles-by -professor-yunus/219-the-problem-of-poverty-in-bangladesh.

Yunus, Muhammad, and Alan Jolis. *Banker to the Poor: Micro-Lending and the Battle against World Poverty*. New York: PublicAffairs, 2003.

Zimmerman, Jamie, Nicole Tosh, and Jamie Holmes. "Investing in Girls: Opportunities for Innovation in Girl-Centered Cash Transfers." Policy Paper. Washington, DC: New America Foundation, 2012.

Žižek, Slavoj. *The Sublime Object of Ideology*. New York: Verso Books, 1989.

Index

Page numbers followed by *f* refer to illustrations.

Girls Discovered project, 125, 127, 128f, 174n8

girls/the Girl: cell phones and, 129; donors, 120; education for, 113–14, 117; financialization of, 9, 129–32; as human capital, 9, 116–17, 118f, 121, 123; investment in, 117–18, 119f, 123–24; liberal feminist aspirations and, 121, 124, 134; risk/opportunities of, 128–29, 132; value or worth of, 114–15; visual representations of, 118, 119f, 120, 121f, 122f

Goldman Sachs, 122–23

governance: experimental, 9; life worth and, 43; NGOization, 87; of population and economy, 61, 141; supranational, 138–39; top-down, 40

governmentality: in Bangladesh, 87; colonial and postcolonial, 70, 82; experimental, 79–80; of family planning, 169n60; Foucault's theory, 149n17; neoliberal, 89, 139, 149n17, 159n19; survey practices and, 64

Grameen Bank, 86, 95, 102, 164n15; global reach of, 88f, 89

gross domestic product (GDP): averted births and, 47, 70; data and calculations for, 22–23, 153n21; fertility reduction and, 114; girls and, 113, 118f, 123, 139; IMF and, 29; invention of, 12; life's contribution to, 6; national income accounts, 26–28, 33; nowcasting, 153n20; per capita, 41–42; as phantasmagrams, 24–25, 27, 54; population planning and, 51; U.S., 22; world distribution map, 27, 28f

gross national product (GNP), 22

Haraway, Donna, 177n14

Hardt, Michael, 161n56

Hartmann, Betsy, 76

Harvard Advisory Group (HAG), 40, 156n22

Hasan, M. A., 166n22

heredity, 3, 11

heterosexual propriety, 69, 117, 141

HIV: care, 79–80; prevention research, 92

Hlanze, Lindi, 175n22

Hoover, Edgar M., 42, 66, 161n52

human capital, 130, 133; concepts of, 115–16; the Girl as, 9, 116–17, 118f, 121, 123

Human Development Index (HDI), 29

humanitarian emergency aid, 101

imaginaries: capitalism, 134; Cold War, 61, 153n25; family planning, 61; feminist, 120; GDP, 24; liberal political, 75, 117; phantasy and phantasma, 25, 153n27; population, 139

immunization, 21, 79, 93, 100

Implications of Worldwide Population Growth for U.S. Security and Overseas Interests (Kissinger), 52, 64, 116

India, 21, 152n7; cholera outbreak, 100; colonial state infrastructure, 39; statisticians and economists, 29; sterilization campaigns, 43, 74

indices, 29–30, 154n37; Maplecroft's, 125–26

industrialization, 37f, 38, 79

inequality, 40, 130

Intel 10x10 campaign, 131f, 175n32

International Centre for Diarrheal Disease Research, Bangladesh (ICDDR,B), 95–96, 98, 102, 171n7. See also Matlab field site

International Crimes Tribunal of Bangladesh, 165n22, 166n24

International Fertility Research Program (IFRP), 92, 169n56

International Monetary Fund (IMF), 29, 41, 138

James, Selma, 31

Kangas, Lenni, 75
Kenny, Katherine, 148n15
Keynes, John Maynard, 43, 152n7; macroeconomy theory, 12, 18, 19f, 20–21
Khan, Ayub, 39, 40, 42, 74
kinship, 177n14
Kissinger, Henry, 52, 64, 86, 116
Kiyo Sue Inui, 147n3
Klein, Melanie, 153n27
knowing, structures of, 7, 62
Knowledge, Attitudes, and Practices (KAP) surveys: family planning, 62–64, 65f, 66, 68; gap issue, 71; importance of affect and, 70–71, 161n51; marketing techniques, 69
Kristof, Nicholas, 124
Kuznets, Simon, 21, 154n31

labor: affective, 161n51; capitalism and, 31, 33, 191n56; cheap/disposable, 40, 90, 94, 123, 133, 140; commodities and, 26; experimental, 91, 168n54; unwaged, 27, 31, 33, 120; value of, 6, 149n16
Lazzarato, Maurizio, 161n50
Leclerc, Georges-Louis, 32
Li, Tania Murray, 159n19
liberal feminism, 121, 124, 134, 139. See also feminists
liberal politics, 26, 75–76, 137, 141
life: abundant, 35, 41–42, 46, 60, 69–70; affirmative, 82, 142, 145; biomedical and biotech research and, 92–93; as capital, 9, 13; capitalism and, 132, 140, 149n16; chances, 6, 46, 80, 81, 104, 141–42, 149n16; cultural, 26; and

death, 2, 140, 143; extraeconomic, 26–27, 33; forms and arrangements of, 141–42, 169n60; infrastructures, 102–4, 141; materializations, 145; politics of, 13–14, 105, 109, 141; reproduction and, 32, 102; subsumption of, 72; unborn, 47–48; value or worth, 3, 6–8, 42–43, 46, 95, 104, 136. See also aggregate life; economization of life; surplus life
Limits to Growth, The (Club of Rome, 1972), 44, 45f
living-beings, 2, 64, 70; reproduction and, 32, 141–42; sexed, 14, 33, 85, 90; transformation of, 13
Lorde, Audre, 81

macroeconomy: atmosphere of, 20, 30–31, 53; data calculations, 21–23; industry and, 138; Keynesian theory of, 12, 18, 19f, 20–21; mass affect and, 71; national, 51, 136; of the nation-state, 6, 27, 28f; phantasmagram of, 129; reproduction and, 33; value and, 148n15
Mahalanabis, Dilip, 171n7
Mahalanobis, Prasanta Chandra, 39
Malthus, Thomas, 148n7; population model, 3–4, 10, 152n7
Mamdani, Mahmood, 71
Maplecroft, 174n4; Girls Discovered project, 125, 127, 128f, 174n8; risk analyses, 125–26
Margaret Sanger Clinic (NY), 66
marketing, 76, 130; research, 62–63, 89, 159n24; social, 69–70, 79, 87
markets, 18, 21, 70, 151n6, 161n51; girls as emerging, 130, 131f, 132, 133; telecommunication, 129
Marx, Karl: formal and real subsumption, 161n56; labor theory of value, 149n16; political economy critique, 25, 31; reproduction theory, 32–33

Marxist feminists, 31, 33, 143

Masco, Joseph, 61, 153n25

material-semiotic actor, 7, 150n20

Matlab field site: Center for Health and Population Research, 101–2; cholera vaccine trials, 98; data collection, 99f, 102, 103f; as a demographic surveillance site, 98–100; establishment and infrastructure, 95–96, 98; map of site, 97f; oral rehydration therapy (ORT), 100–101, 171n7; scientists, 96

Mbembe, Achille, 84

McClintock, Anne, 11

Mills, Charles, 26

miscarriages, 48

Mitchell, Timothy, 148n15

mobile phones, 129–31

modernization, 70; in Bangladesh, 79; demographic transition and, 36, 37f, 38, 136; in Pakistan, 41

Moore, Hugh, 154n5, 155n6

Morgan, Lewis Henry, 148n9

mortality rates. See death rates

Mosley, Henry, 170n1

Multidimensional Poverty Index (MPI), 30

National Institutes of Health (NIH), 101, 170n2

nation-state, 1, 14, 18, 31, 129, 135–37; macroeconomy of, 6, 27, 28f, 72, 143

Nayakrishi Andolon (the new cultivation movement), 105–7, 109, 143, 145

necropolitics, 84, 103, 141

Negri, Antonio, 161n56

neoliberalism: anticipation and, 114; economics and, 7, 149n17; experimentality and, 89–90; governmentality, 89, 139, 149n17, 159n19; invention in Bangladesh, 9, 87, 90

NGOization, 87, 90

NGOs, 52, 60, 91; Bangladesh, 87; feminist, 120, 121. See also BRAC (Bangladesh Rehabilitation and Assistance Committee)

Nguyen, Vinh-Kim, 79

Nike Foundation, 117, 122–23, 173n9; Maplecroft's analyses for, 125, 126, 127, 174n7

Notestein, Frank, 162n15; on "Abundant Life," 35, 41; as consultant to Pakistan, 39–41; demographic transition model, 36, 38, 42, 155n9

organic farming movement, 105–8

Orr, Jackie, 61, 153n25

overpopulation, 12, 135; in Bangladesh, 85; fear and panic of, 44, 46, 52, 64, 137; Malthus's concern for, 4, 10, 152n7

Pakistan: economic development, 40–41; family planning programs, 74; independence war, 83, 165n22; PL480 money, 170n2; state infrastructure and geography, 39–40, 82–83; statisticians and economists, 39, 41

Papanek, Gustav, 40

Patel, Geeta, 69, 116

Pateman, Carole, 26

Pearl, Raymond, 148n8; eugenics and, 3, 11, 147n6; fruit fly experiments, xf, 1–2, 77; population research, 3–5, 9–10; s-curve graphs, 2f, 4, 5f, 10–13, 36, 38

phantasmagrams: averted births as, 48–49; capitalism as, 134; Cold War and, 153n25; demographic transition models as, 50–53; economization of life as, 62; GDP as, 24–25, 27, 54; the Girl as, 117, 120, 122, 125, 127, 134; macroeconomy as, 129; population and economy as, 140; technoscience as, 56 phantasy, 24, 74, 137, 144; big data's, 127; quantification and, 134; technoscience and, 52, 56; term usage, 153n27

pharmaceutical research, 91–92, 168n54

political economy, 3; Marx's critique, 25, 31; natural history and, 32, 148n7

populate, definition, 135

population: distributed reproduction and, 141; economy and, 1, 3, 6–7, 13–14, 36, 108–9; governance of, 61; imaginary, 139; Malthus *vs.* Pearl's model of, 4; optimum, 10, 150n24; politics against, 137, 139, 141, 144–45; pressure, 35, 85; problem of, 2, 8, 10–12, 36, 44, 61, 135–39; quantitative and experimental approaches to, 3–4, 9–10; race and, 6, 135; scholarship on, 150n21; term usage, 135–37, 176n1, 176n7. *See also* demographic transition models

population bomb, 36, 136, 137; term origin, 154n5

Population Bomb (Ehrlich), 44, 155n5, 159n24

population control, 6, 155n6, 159n24; in Bangladesh, 78, 108; in China, 43; in Cold War period, 35–36; computer simulations on, 50; in Pakistan, 39, 41, 74; racism and, 138; unborn babies and, 49, 93, 101; U.S. funding for, 78, 150n30

Population Council, 38, 44, 62, 76, 159n30

population growth: of Algerians, 4–5; in biology of fruit flies, x*f*, 1–2; climate change and, 137–38; economic modernization and, 36, 37*f*, 38; fear of, 43–44; Pearl's s-curve, 2*f*, 5*f*, 36, 38, 147n1; reduction in, 35, 41–42, 45*f*, 52, 75, 157n14

Population Services International (PSI), 69, 87, 160n49; global reach of, 88*f*, 89

postcolonialism: Cold War and, 9, 13, 150n23; living laboratories and, 41, 156n25; modernization projects, 79; neoliberalism, 90; population studies of South Asia, 64, 160n33; surveys and, 62–63

postcolonial thick data, 9, 59, 104, 114, 137; description, 158n7; at Matlab, 102, 103*f*; of women and girls, 127, 128*f*

poverty: in Bangladesh, 90; Communism and, 35; distributions of, 136; GDP and, 28; of girls, 120, 131; indices of, 30; of women, 31

precarity: in Bangladesh, 90–91; distributions of, 136; of experimentality, 86, 89–90; of life and death, 93–94, 102–4, 135, 145

pregnancy, 48–49; unwanted, 67*f*

quantification practices, 6–7, 9, 20, 24, 53; surveys, 62–64, 65*f*, 66. *See also* data collection; demographic transition models; gross domestic product (GDP)

Quesnay, François, 31

Qureshi, Moeen, 41

racialization: of bodies, 11–12, 168n54; evolution and, 11, 13; heredity and, 3; of life, 6, 7, 104

racism: biology and, 3, 148n6; bodily difference and, 12; colonialism and, 31; distribution of life chances and, 104; history of, 6; infrastructures and, 139–40; of Keynes, 152n7; of Pearl, 10; population control and, 138, 155n6; quantitative models and, 46; state, 41

Rao, V. K. R. V., 39

rape, 83, 85, 126, 166n22

Ravenholt, Reimert, 67–68

reproduction: assisted, 169n60; capital-

ism and, 7, 33; Cold War, 35; definition of, 31–32; distributed, 9, 141–44,
169n60; economy and, 9, 31; experiments, 91; feminists/feminism and,
13, 139, 142, 150n32; human and nonhuman, 108; infrastructures of, 139–
40; Marx's theory, 32–33; politics of,
8, 109, 142; stratified, 177n14; unwaged
labor and, 31
reproductive justice, 142–43
research infrastructures, 71–72, 92; at
Matlab, 98–99, 101, 103
Rheinberger, Hans-Jörg, 80
rice, 106f, 107f, 108
risk analyses, 125–27, 134
Rogers, Everett, 74
Roquiah, Begum, 55–56, 105

Sanger, Margaret, 2
Schabas, Margaret, 148n7
Schultz, Theodore, 115–16
Schwittay, Anke, 175n25
science and technology studies (STS), 13,
150n20, 152n6
seed saving, 106f, 107
self-help projects, 78, 82, 164n15
Sen, Amartya, 29, 84
Sexton, Jared, 176n2
Shoaib, Muhammad, 41
Simon, Julian, 63, 159n24
Sloterdijk, Peter, 151n3
Smith, Adam, 31, 148n7
Social Marketing Project (SMP) of Bangladesh, 89, 160n49
social science practices: affective atmospheres and, 72; counting, 61; economization of life and, 6; quantification
and modeling, 20, 24, 28, 40; in South
Asia, 39–41, 78; surveying, 63, 66; between U.S. and Bangladesh, 7, 8–9
state: economic planning, 43; fiscal poli-

cies, 18, 20; governmentality, 149n17;
investment in children, 42
states of expectation, 20, 22
sterilization, 89; averted births from, 48,
49; campaign in Bangladesh, 76, 79;
campaign in India, 43, 74; minimization of supports for, 76–77, 163n21;
payoffs for, 75
Strathern, Marilyn, 177n14
straw, 107f, 108
subsumption, 72, 81, 140, 161n56
Sultana's Dream (Roquiah), 55–56, 105
Summers, Lawrence, 113, 115, 171n9, 173n1
supply chain: capitalism, 143; global, 90,
94, 125; logistics, 9, 126
surplus life, 51, 61, 70, 135, 138, 161n50; calculations of, 9, 93; precarious life as,
93, 145; race and, 12, 104, 139
surplus value, 6, 25, 33

Tadiar, Neferti, 149n16, 176n33
Technical Military Planning Operation
(TEMPO), 50, 157n14
technoscience, 168n54, 169n60; dreams,
32, 51, 53–54, 55–56; feminist, 56, 121,
177n14
Thompson, Charis, 92
time or temporality, 12, 20–21; anticipation and, 114; in demographic transition models, 36, 37f, 42–43

UBINIG (Unnayan Bikalper Nitinirdharoni Gobeshona, or Policy Research for Development Alternative),
143, 165n16; experimental farm, 106f,
107f, 107–8; philosophies and practices, 105–7
Ul Haq, Mahbub, 29
unborn lives. *See* averted births
United Nations (UN): Fund for Population Activities, 76; GDP development

United Nations (*continued*)
standards, 28; Global Pulse project, 129; Millennium Development Goals, 30

United States: cholera research, 95–96, 170n2; Cold War military projects, 35, 61, 95, 153n25; data collection, 59; economy, 22; family planning projects, 36, 47, 63–64, 150n23; foreign aid, 13, 35, 41, 104, 150n30; foreign policy, 64, 85, 101; GDP and GNP, 22, 24; imperialism, 78; national income account (1934), 21; newspaper headlines, 17; scientists and researchers, 40–41, 96, 98, 102, 170n2; social science practices, 8–9

United States Agency for International Development (USAID): approach to contraception, 68, 160n43; family planning funding, 50, 52, 63, 66, 85, 155n7; founding, 35; incentive programs, 75–76; Matlab funding, 101; policies and objectives, 51, 67, 157n14; research contracts, 92

value, regime of, 5–6, 149n15
vasectomies, 73–74
violence, 52, 93, 135, 139, 145; in Bangladeshi history, 79, 83–84; of capital, 141; racial, 12; sexual, 85, 143
voluntary organizations, 78
vulnerability, 104, 127

Waldby, Catherine, 168n54
war crimes, 165n22, 166n24
Warhurst, Alyson, 174n4
Waring, Marilyn, 153n21
waste-making, 138–39, 141
women: averted births and, 49–50; contraceptive choices, 43; data collection on, 59–60; independence, 43; infertility, 48; labor of, 31, 33, 90, 94; modernity of, 62; Muslim, 55–56, 67, 105; as research subjects, 82, 92, 105, 165n16; scientists, 55; violence toward, 85. *See also* feminists; girls/the Girl
World Bank, 41, 76, 86, 138; funding, 42, 85; GDP distribution map, 28*f*
World Fertility Survey, 63
World Population Conference (1927), 2–3, 147n3
Wright, Melissa, 94

Yellen, Janet, 30*f*
Yunus, Muhammad, 86–87